Greek Biology and Greek Medicine

Charles Joseph Singer

Writat

This edition published in 2020

ISBN : 9789390439171

Design and Setting By
Writat
(An Imprint of Vij Publishiing Group)
www.vijbooks.com
contact@vijpublishing.com

PREFACE 1

GREEK BIOLOGY 3
    § 1. *Before Aristotle* 3
    § 2. *Aristotle* 19
    § 3. *After Aristotle* 56

GREEK MEDICINE 81

PREFACE

This little book is an attempt to compress into a few pages an account of the general evolution of Greek biological and medical knowledge. The section on *Aristotle* appears here for the first time. The remaining sections are reprinted from articles contributed to a volume *The Legacy of Greece* edited by Mr. R. W. Livingstone, the only changes being the correction of a few errors and the addition of some further references to the literature.

In quoting from the great Aristotelian biological treatises, the *History of Animals*, the *Parts of Animals*, and the *Generation of Animals*, I have usually availed myself of the text of the Oxford translation edited by Mr. W. D. Ross. For the *De anima* I have used the version of Mr. R. D. Hicks.

I have to thank my friends Mr. R. W. Livingstone, Dr. E. T. Withington, and Mr. J. D. Beazley for a number of suggestions. To my colleague Professor Arthur Platt I have to record my gratitude not only for much help in the writing of these chapters but also for his kindness and patience in reading and rereading the work both in manuscript and proof. I am specially indebted, moreover, to the notes appended to his translation of the *Generation of Animals*.

C. S.

UNIVERSITY COLLEGE, LONDON.
March 1922.

GREEK BIOLOGY

§ 1. *Before Aristotle*

What is science? It is a question that cannot be answered easily, nor perhaps answered at all. None of the definitions seem to cover the field exactly; they are either too wide or too narrow. But we can see science in its growth and we can say that being a process it can exist only as growth. Where does the science of biology begin? Again we cannot say, but we can watch its evolution and its progress. Among the Greeks the accurate observation of living forms, which is at least one of the essentials of biological science, goes back very far. The word *Biology*, used in our sense, would, it is true, have been an impossibility among them, for *bios* refers to the life of man and could not be applied, except in a strained or metaphorical sense, to that of other living things.[1] But the *ideas* we associate with the word are clearly developed in Greek philosophy and the foundations of biology are of great antiquity.

The Greek people had many roots, racial, cultural, and spiritual, and from them all they inherited various powers and qualities and derived various ideas and traditions. The most suggestive source for our purpose is that of the Minoan race whom they dispossessed and whose lands they occupied. That highly gifted people exhibited in all stages of its development a marvellous power of graphically representing animal forms, of which the famous Cretan friezes, Vaphio cups (<u>Fig. 5</u>), and Mycenean lions provide well-known examples. It is difficult not to believe that the Minoan element, entering into the mosaic of peoples that we call the Greeks, was in part at least responsible for the like graphic power developed in the Hellenic world, though little contact has yet been demonstrated between Minoan and archaic Greek Art.

For the earliest biological achievements of Greek peoples we have to rely largely on information gleaned from artistic remains. It is true that we have a few fragments of the works of both Ionian and Italo-Sicilian philosophers, and in them we read of theoretical speculation as to the nature of life and of the soul, and we can thus form some idea of the first attempts of such workers as Alcmaeon of Croton (*c.* 500 B. C.) to lay bare the structure of animals by dissection.[2] The pharmacopœia also of some of the earliest works of the Hippocratic collection betrays considerable knowledge of both native and foreign plants.[3] Moreover, scattered through the pages of Herodotus and other early writers is a good deal of casual information concerning animals and plants, though such material is second-hand and gives us little information concerning the habit of exact observation that is the necessary basis of science.

FIG. 1. Lioness and young from an Ionian vase of the sixth century B. C. found at Caere in Southern Etruria (Louvre, Salle E, No. 298), from *Le Dessin des Animaux en Grèce d'après les vases peints*, by J. Morin, Paris (Renouard), 1911. The animal is drawing itself up to attack its hunters. The scanty mane, the form of the paws, the udders, and the dentition are all heavily though accurately represented.

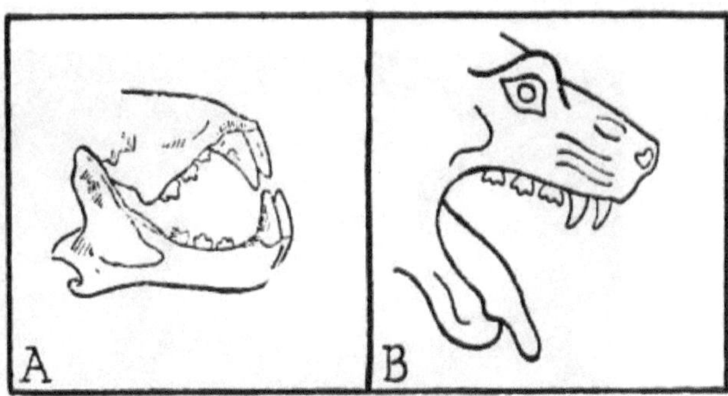

FIG. 2. A, Jaw bones of lion; B, head of lioness from Caere vase (Fig. 1), after Morin. Note the careful way in which the artist has distinguished the molar from the cutting teeth.

Something more is, however, revealed by early Greek Art. We are in possession of a series of vases of the seventh and sixth centuries before the Christian era showing a closeness of observation of animal forms that tells of a people awake to the study of nature. We have thus portrayed for us a number of animals—plants seldom or never appear—and among the best rendered are wild creatures; we see antelopes quietly feeding or startled at a sound, birds flying or picking worms from the ground, fallow deer forcing their way through thickets, browsing peacefully, or galloping away, boars facing the hounds and dogs chasing hares, wild cattle forming their defensive circle, hawks seizing their prey. Many of these exhibit minutely accurate observation. The very direction of the hairs on the animals' coats has sometimes been closely studied, and often the muscles are well rendered. In some cases even the dentition has been found accurately portrayed, as in a sixth century representation on an Ionian vase of a lioness—an animal then very rare on the Eastern Mediterranean littoral, though still well known in Babylonia, Syria, and Asia Minor. The

details of the work show that the artist must have examined
the animal in captivity (Figs. <u>1</u> and <u>2</u>).

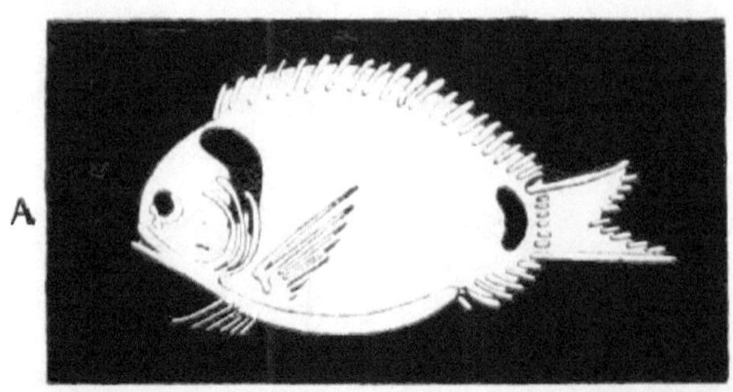

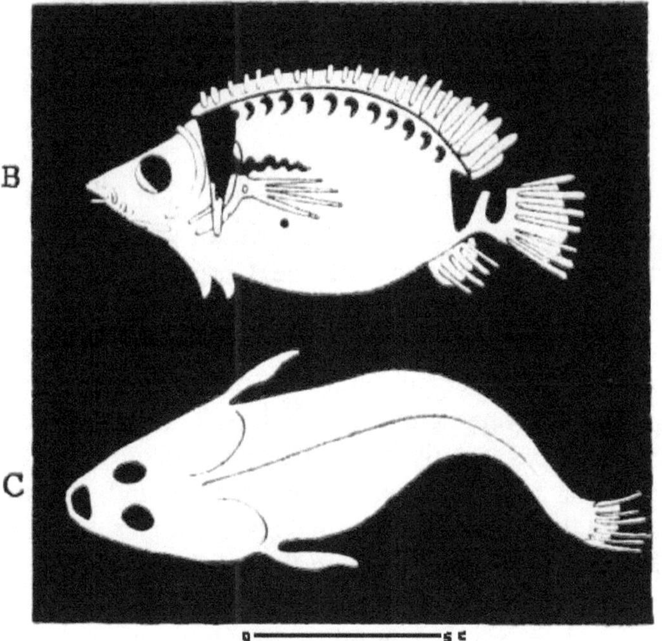

FIG. 3. Paintings of fish on plates.
Italo-Greek work of the fourth century B.
C. From Morin.

A.   Sargus vulgaris.
B.   Crenilabrus mediterraneus.
C.   Uranoscopus scaber?

Animal paintings of this order are found scattered over the Greek world with special centres or schools in such places as Cyprus, Boeotia, or Chalcis. The very name for a painter in Greek, *zoographos*, recalls the attention paid to living forms. By the fifth century, in representing them as in other departments of Art, the supremacy of Attica had asserted itself, and there are many beautiful Attic vase-paintings of animals to place by the side of the magnificent horses' heads of the Parthenon (Fig. 6). In Attica, too, was early developed a characteristic and closely accurate type of representation of marine forms, and this attained a wider vogue in Southern Italy in the fourth century. From the latter period a number of dishes and vases have come down to us bearing a large variety of fish forms, portrayed with an exactness that is interesting in view of the attention to marine creatures in the surviving literature of Aristotelian origin (Fig. 3).

These artistic products are more than a mere reflex of the daily life of the people. The habits and positions of animals are observed by the hunter, as are the forms and colours of fish by the fisherman; but the methods of huntsman and fisher do not account for the accurate portrayal of a lion's dentition, the correct numbering of a fish's scales or the close study of the lie of the feathers on the head, and the pads on the feet, of a bird of prey (Fig. 4). With observations such as these we are in the presence of something worthy of the name *Biology*. Though but little literature on that topic earlier than the writings of Aristotle has come down to us, yet both the character of his writings and such paintings and pictures as these, suggest the existence of a strong interest and a wide literature, biological in the modern sense, antecedent to the fourth century.

FIG. 4. Head and talons of the Sea-eagle, *Haliaëtus albicilla*:

A,　From an Ionic vase of the sixth century B. C.
B,　Drawn from the object.
From Morin.

Greek science, however, exhibits throughout its history a peculiar characteristic differentiating it from the modern scientific standpoint. Most of the work of the Greek scientist was done in relation to man. Nature interested him mainly in relation to himself. The Greek scientific and philosophic world was an anthropocentric world, and this comes out in the overwhelming mass of medical as distinct from biological writings that have come down to us. Such, too, is the sentiment expressed by the poets in their descriptions of the animal creation:

Many wonders there be, but naught more wondrous than man:

The light-witted birds of the air, the beasts of the weald and the wood He traps with his woven snare, and the brood of the briny flood. Master of cunning he: the savage bull, and the hart Who roams the mountain free, are tamed by his infinite art. And the shaggy rough-maned steed is broken to bear the bit.

Sophocles, *Antigone*, verses 342 ff. (Translation of F. Storr.)

It is thus not surprising that our first systematic treatment of animals is in a practical medical work, the περὶ διαίτης, *On regimen*, of the Hippocratic Collection. This very peculiar treatise dates from the later part of the fifth century. It is strongly under the influence of Heracleitus (*c.* 540-475) and contains many points of view which reappear in later philosophy. All animals, according to it, are formed of fire and water, nothing is born and nothing dies, but there is a perpetual and eternal revolution of things, so that change itself is the only reality. Man's nature is but a parallel to that of the universal nature, and the arts of man are but an imitation or reflex of the natural arts or, again, of the bodily functions. The soul, a mixture of water and fire, consumes itself in infancy and old age, and increases during adult life. Here, too, we meet with that singular doctrine, not without bearing on the course of later biological thought, that in the foetus all parts are formed simultaneously. On the proportion of fire and water in the body all depends, sex, temper, temperament, intellect. Such speculative ideas separate this book from the sober method of the more typical Hippocratic medical works with which indeed it has little in common.

After having discussed these theoretical matters the work turns to its own practical concerns, and in the course of setting out the natures of foods gives in effect a rough classification of animals. These are set forth in groups, and from among the larger groups only the reptiles and insects are missing. The list has been described, perhaps hardly with justification, as the *Coan classificatory system*. We have here, indeed, no *system* in the sense in which that word is now applied to the animal kingdom, but we have yet some sort of definite arrangement of animals according to their supposed natures. The passage opens with mammals, which are divided into domesticated and wild, the latter being mentioned in order according to size, next follow the land-birds, then the water-fowl, and then the fishes. These fish are divided into (1) the haunters of the shore, (2) the free-swimming forms, (3) the cartilaginous fishes or Selachii, which are not so named but are placed together, (4) the mud-loving forms, and (5) the fresh-water fish. Finally come invertebrates arranged in some sort of order according to their structure. The characteristic feature of the 'classification' is the separation of the fish from the remaining vertebrates and of the invertebrates from both. Of the fifty animals named no less than twenty are fish, about a fifth of the number studied by Aristotle, but we must remember that here only edible species are mentioned. The existence of the work shows at least that in the fifth century there was already a close and accurate study of animal forms, a study that may justly be called scientific. The predominance of fish and their classification in greater detail than the other groups is not an unexpected feature. The Mediterranean is especially rich in these forms, the Greeks were a maritime people, and Greek literature is full of imagery drawn from the fisher's craft. From Minoan to Byzantine times the variety, beauty, and colour of fish made a deep impression on Greek minds as reflected in their art.

Much more important however for subsequent biological development, than such observations on the nature and habits of animals, is the service that the Hippocratic physicians rendered to Anatomy and to Physiology, departments in which the structure of man and of the domesticated animals stands apart from that of the rest of the animal kingdom. It is with the nature and constitution of man that most of the surviving early biological writings are concerned, and in these departments are unmistakable tendencies towards systematic arrangement of the material. Thus we have division and description of the body in sevens from the periphery to the centre and from the vertex to the sole of the foot,[4] or a division into four regions or zones.[5] The teaching concerning the four elements and four humours too became of great importance and some of it was later adopted by Aristotle. We also meet numerous mechanical explanations of bodily structures, comparisons between anatomical conditions encountered in related animals, experiments on living creatures,[6] systematic incubation of hen's eggs for the study of their development, parallels drawn between the development of plants and of human and animal embryos, theories of generation, among which is that which was afterwards called 'pangenesis'— discussion of the survival of the stronger over the weaker— almost our survival of the fittest—and a theory of inheritance of acquired characters.[7] All these things show not only extensive knowledge but also an attempt to apply such knowledge to human needs. When we consider how even in later centuries biology was linked with medicine, and how powerful and fundamental was the influence of the Hippocratic writings, not only on their immediate successors in antiquity, but also on the Middle Ages and right into the nineteenth century, we shall recognize the significance of these developments.

Fig. 5. MINOAN GOLD CUP. SIXTEENTH CENTURY
B. C.

Fig. 6. HORSE'S HEAD. FROM PARTHENON. 440 B. C.

Such was the character of biological thought within the fifth century, and a generation inspired by this movement produced some noteworthy works in the period which immediately followed. In the treatise περὶ τροφῆς, *On nourishment,* which may perhaps be dated about 400 B. C., we learn of the pulse for the first time in Greek medical literature, and read of a physiological system which lasted until the time of Harvey, with the arteries arising from the heart and the veins from the liver. Of about the same date is a work περὶ καρδίης, *On the heart,* which describes the ventricles as well as the great vessels and their valves, and compares the heart of animals with that of man.

A little later, perhaps 390 B. C., is the treatise περὶ σαρκῶν, *On muscles,* which contains much more than its title suggests. It has the old system of sevens and, inspired perhaps by the philosophy of Heracleitus (*c.* 540-475), describes the heart as sending air, fire, and movement to the different parts of the body through the vessels which are themselves constantly in movement. The infant in its mother's womb is believed to draw in air and fire through its mouth and to eat *in utero.* The action of air on the blood is compared to its action on fire. In contrast to some of the other Hippocratic treatises the central nervous system is in the background; much attention, however, is given to the special senses. The brain resounds during audition. The olfactory nerves are hollow, lead to the brain, and convey volatile substances to it which cause it to secrete mucus. The eyes also have been examined, and their coats and humours roughly described; an allusion, the first in literature, is perhaps made to the crystalline lens, and the eyes of animals are compared with those of man. There is evidence not only of dissection but of experiment, and in efforts to compare the resistance of various tissues to such processes as boiling, we may see the small beginning of chemical physiology.

An abler work than any of these, but exhibiting less power of observation is a treatise, περὶ γονῆς, *On generation*, that may perhaps be dated about 380 B. C.[8] It exhibits a writer of much philosophic power, very anxious for physiological explanations, but hampered by ignorance of physics. He has, in fact, the weaknesses and in a minor degree the strength of his successor Aristotle, of whose great work on generation he gives us a foretaste. He sets forth in considerable detail a doctrine of pangenesis, not wholly unlike that of Darwin. In order to explain the phenomena of inheritance he supposes that vessels reach the seed, carrying with them samples from all parts of the body. He believes that channels pass from all the organs to the brain and then to the spinal marrow (or to the marrow direct), thence to the kidneys and on to the genital organs; he believes, too, that he knows the actual location of one such channel, for he observes, wrongly, that incision behind the ears, by interrupting the passage, leads to impotence. As an outcome of this theory he is prepared to accept inheritance of acquired characters. The embryo develops and breathes by material transmitted from the mother through the umbilical cord. We encounter here also a very detailed description of a specimen of exfoliated *membrana mucosa uteri* which our author mistakes for an embryo, but his remarks at least exhibit the most eager curiosity.[9]

The author of this work on generation is thus a 'biologist' in the modern sense, and among the passages exhibiting him in this light is his comparison of the human embryo with the chick. 'The embryo is in a membrane in the centre of which is the navel through which it draws and gives its breath, and the membranes arise from the umbilical cord.... The structure of the child you will find from first to last as I have already described.... If you wish, try this experiment: take twenty or more eggs and let them be incubated by two or more hens. Then each day from the second to that of hatching remove an egg, break it, and examine it. You will find exactly as I say, for the nature of the bird can be likened to that of man. The membranes [you will see] proceed from the umbilical cord, and all that I have said on the subject of the infant you will find in a bird's egg, and one who has made these observations will be surprised to find an umbilical cord in a bird's egg.'[10]

The same interest that he exhibits for the development of man and animals he shows also for plants.

'A seed laid in the ground fills itself with the juices there contained, for the soil contains in itself juices of every nature for the nourishment of plants. Thus filled with juice the seed is distended and swells, and thereby the power (= faculty ἡ δῦύναμις) diffused in the seed is compressed by living principle (pneuma) and juice, and bursting the seed becomes the first leaves. But a time comes when these leaves can no longer get nourished from the juices in the seed. Then the seed and the leaves erupt, for urged by the leaves the seed sends down that part of its power which is yet concentrated within it and so the roots are produced as an extension of the leaves. When at last the plant is well rooted below and is drawing its nutriment from the earth, then the whole grain disappears, being absorbed, save for the husk, which is the most solid part; and even that, decomposing in the earth, ultimately becomes invisible. In time some of the leaves put forth branches. The plant being thus produced by humidity from the seed is still soft and moist. Growing actively both above and below, it cannot as yet bear fruit, for it has not the quality of force and reserve (δύναμις ἰσχυρὴ καὶ πιαρά) from which a seed can be precipitated. But when, with time, the plant becomes firmer and better rooted, it develops veins as passages both upwards and downwards, and it draws from the soil not only water but more abundantly also substances that are denser and fatter. Warmed, too, by the sun, these act as a ferment to the extremities and give rise to fruit after its kind. The fruit thus develops much from little, for every plant draws from the earth a power more abundant than that with which it started, and the fermentation takes place not at one place but at many.'[11]

Nor does our author hesitate to draw an analogy between the plant and the mammalian embryo. 'In the same way the infant lives within its mother's womb and in a state corresponding to the health of the mother ... and you will find a complete similitude between the products of the soil and the products of the womb.'

The early Greek literature is so scantily provided with illustrations drawn from botanical study, that it is worth considering the remarkable comparison of generation of plants from cuttings with that from seeds in the same work.

> 'As regards plants generated from cuttings ... that part of a branch where it was cut from a tree is placed in the earth and there rootlets are sent out. This is how it happens: The part of the plant within the soil draws up juices, swells, and develops a *pneuma* (πνεῦμα ἴσχει), but not so the part without. The pneuma and the juice concentrate the power of the plant below so that it becomes denser. Then the lower end erupts and gives forth tender roots. Then the plant, taking from below, draws juices from the roots and transmits them to the part above the soil which thus also swells and develops pneuma; thus the power from being diffused in the plant becomes concentrated and budding, gives forth leaves.... Cuttings, then, differ from seeds. With a seed the leaves are borne first, then the roots are sent down; with a cutting the roots form first and then the leaves.'[12]

But with these works of the early part of the fourth century the first stage of Greek biology reaches its finest development. Later Hippocratic treatises which deal with physiological topics are on a lower plane, and we must seek some external cause for the failure. Nor have we far to seek. This period saw the rise of a movement that had the most profound influence on every department of thought. We see the advent into the Greek world of a great intellectual movement as a result of which the department of philosophy that dealt with nature receded before Ethics. Of that intellectual revolution—perhaps the greatest the world has seen—Athens was the site and Socrates (470-399) the protagonist. With the movement itself and its characteristic fruit we are not concerned. But the great successor and pupil of its founder gives us in the *Timaeus* a picture of the depth to which natural science can be degraded in the effort to give a specific teleological meaning to all parts of the visible Universe. The book and the picture which it draws, dark and repulsive to the mind trained in modern scientific method, enthralled the imagination of a large part of mankind for wellnigh two thousand years. Organic nature appears in this work of Plato (427-347) as the degeneration of man whom the Creator has made most perfect. The school that held this view ultimately decayed as a result of its failure to advance positive knowledge. As the centuries went by its views became further and further divorced from phenomena, and the bizarre developments of later Neoplatonism stand to this day as a warning against any system which shall neglect the investigation of nature. But in its decay Platonism dragged science down and destroyed by neglect nearly all earlier biological material. Mathematics, not being a phenomenal study, suited better the Neoplatonic mood and continued to advance, carrying astronomy with it for a while—astronomy that affected the life of man and that soon became the handmaid of astrology; medicine, too, that determined the conditions of man's life, was also cherished, though often mistakenly, but pure science was doomed.

But though the ethical view of nature overwhelmed science in the end, the advent of the mighty figure of Aristotle (384-322) stayed the tide for a time. Yet the writer on Greek Biology remains at a disadvantage in contrast with the Historian of Greek Mathematics, of Greek Astronomy, or of Greek Medicine, in the scantiness of the materials for presenting an account of the development of his studies before Aristotle. The huge form of that magnificent naturalist completely overshadows Greek as it does much of later Biology.

§ 2. *Aristotle*

With Aristotle we come in sight of the first clearly defined personality in the course of the development of Greek biological thought—for the attribution of the authorship of the earlier Hippocratic writings is more than doubtful, while the personality of the great man by whose name they are called cannot be provided with those clear outlines that historical treatment demands.

Aristotle was born in 384 B. C. at Stagira, a Greek colony in the Chalcidice a few miles from the northern limit of the present monastic settlement of Mount Athos. His father, Nicomachus, was physician to Amyntas III of Macedonia and a member of the guild or family of the Asclepiadae. From Nicomachus he may have inherited his taste for biological investigation and acquired some of his methods. At seventeen Aristotle became a pupil of Plato at Athens. After Plato's death in 347 Aristotle crossed the Aegean to reside at the court of Hermias, despot of Atarneus in Mysia, whose niece, Pythias, he married. It is not improbable that the first draft of Aristotle's biological works and the mass of his own observations were made during his stay in this region, for in his biological writings much attention is concentrated on the natural history of the Island of Lesbos, or Mytilene, that lies close opposite to Atarneus. Investigation has shown that in the *History of Animals* there are frequent references to places on the northern and eastern littoral of the Aegean, and especially to localities in the Island of Lesbos; on the other hand places in Greece proper are but seldom mentioned.[13] Thus his biological investigations, in outline at least, are probably the earliest of his extant works and preceded the philosophical writings which almost certainly date from his second sojourn in Athens.

Fig. 7.   ARISTOTLE

From HERCULANEUM
Probably work of fourth century B. C.

In 342 B. C., at the request of Philip of Macedon, Aristotle became tutor to Philip's son, Alexander. He remained in Macedonia for seven years and about 336, when

Alexander departed for the invasion of Asia, returned to Athens where he taught at the Lyceum and established his famous school afterwards called the Peripatetic. Most of his works were produced during this the closing period of his life between 335 and 323 B. C. After Alexander's death in 323 and the break up of his empire, Aristotle, who was regarded as friendly to the Macedonian power, was placed in a difficult position. Regarded with enmity by the anti-Macedonian party, he withdrew from Athens and died soon after in 322 B. C. at Chalcis in Euboea at about sixty-two years of age.

The scientific works to which Aristotle's name is attached may be divided into three groups, physical, biological, and psychological. In size they vary from such a large treatise as the *History of Animals* to the tiny tracts which go to make up the *Parva naturalia*. So far as the scientific writings can be distinguished as separate works they may be set forth as follows:

Physics.

φυσικὴ ἀκρόασις	*Physics.*
περὶ γενέσεως καὶ φθορᾶς	*On coming into being and passing away.*
περὶ οὐρανοῦ.	*On the heavens.*
μετεωρολογικά.	*Meteorology.*
[περὶ κόσμου.	*On the universe.*]
[μηχανικά.	*Mechanics.*]
[περὶ ἀτόμων γραμμῶν.	*On indivisible lines.*]
[ἀνέμων θέσεις καὶ προσηγορίαι.	*Positions and descriptions of winds.*]

Biology in the restricted sense.
(a) *Natural History.*

περὶ τὰ ζῷα ἱστορίαι.	*Inquiry about animals = Historia animalium.*

περὶ ζῴων μορίων.   *On parts of animals.*
περὶ ζῴων γενέσεως.   *On generation of animals.*
[περὶ φυτῶν.   *On plants.*]

(b) *Physiology.*
περὶ ζῴων πορείας.   *On progressive motion of animals.*

περὶ μακροβιότητος καὶ βραχυβιότητος.   *On length and shortness of life.*
περὶ ἀναπνοῆς.   *On respiration.*
περὶ νεότητος καὶ γήρως.   *On youth and age.*
[περὶ ζῴων κινήσεως.   *On motion of animals.*]
[φυσιογνωμονικά.   *On physiognomy.*]
[περὶ πνεύματος.   *On innate spirit.*]

Psychology and Philosophy with biological bearing.
περὶ ψυχῆς.   *On soul.*
περὶ αἰσθήσεως καὶ αἰσθητῶν.   *On sense and objects of sense.*
περὶ ζωῆς καὶ θανάτου.   *On life and death.*
περὶ μνήμης καὶ ἀναμνήσεως.   *On memory and reminiscence.*
περὶ ὕπνου καὶ ἐγρηγόρσεως.   *On sleep and waking.*

περὶ ἐνυπνίων.   *On dreams.*
[προβλήματα.   *Problems.*]
[περὶ χρωμάτων.   *On colours.*]
[περὶ ἀκουστῶν.   *On sounds.*]
[περὶ τῆς καθ' ὕπνον μαντικῆς.   *On prophecy in sleep.*]

Of these works some, the names of which are placed here in brackets, are clearly spurious in that they were neither written by Aristotle nor are they in any form approaching that in which they were cast by him. Yet all are of very considerable antiquity and contain fragments of his tradition in a state of greater or less corruption. In addition to works here enumerated there are many others which are spurious in a yet further sense in that they are merely fathered on Aristotle and contain no trace of his spirit or method. Such, for example, is the famous mediaeval work of oriental origin known as the *Epistle of Aristotle to Alexander*.

In a general way it may be stated that the *physical* works, with which we are not here directly concerned, while they show ingenuity, learning, and philosophical power, yet betray very little direct and original observation. They have exerted enormous influence in the past and for at least two thousand years provided the usual physical conceptions of the civilized world both East and West. After the Galilean revolution in physics, however, they became less regarded and they are not now highly esteemed by men of science. The *biological* works of Aristotle, on the other hand, excited comparatively little interest during the Middle Ages, but from the sixteenth century on they have been very closely studied by naturalists. From the beginning of the nineteenth century, and especially as a result of the work of Cuvier, Richard Owen, and Johannes Müller, Aristotle's reputation as a naturalist has risen steadily, and he is now universally admitted to have been one of the very greatest investigators of living nature.

The philosophical bases of Aristotle's biology are mainly to be found in the treatise *On soul* and in that *On the generation of animals*. His actual observations are contained in this latter work—which is in many ways his finest scientific production—in the great collection on the *History of animals*, and in the remarkable treatise *On parts of animals*. Certain of his deductions concerning the nature and mechanism of life can be found in his two works which deal with the movements of animals (one of which is very doubtfully genuine) and in his tracts *On respiration, On sleep*, &c. The treatise *On plants* and the *Problems* in their present form are late and spurious, but they are based on works of members of his school. They were, however, perhaps originally prepared at the other end of the Greek world in Magna Graecia.

Aristotle was a most voluminous author and his biological writings form but a small fraction of those to which his name is attached. Yet these biological works contain a prodigious number of first-hand observations and it has always been difficult to understand how one investigator could collect all these facts, however rapid his work and skilful his methods. The explanations that have reached us from antiquity are, indeed, picturesque, but they are neither credible in themselves nor are they consistent with each other. Thus Pliny writing about A. D. 77 says 'Alexander the Great, fired by desire to learn of the natures of animals, entrusted the prosecution of this design to Aristotle.... For this end he placed at his disposal some thousands of men in every part of Asia and Greece, and among them hunters, fowlers, fishers, park-keepers, herds-men, bee-wards, as well as keepers of fish-ponds and aviaries in order that no creature might escape his notice. Through the information thus collected he was able to compose some fifty volumes.'[14] Athenaeus, who lived in the early part of the third century A. D., assures us that 'Aristotle the Stagirite received eight hundred talents [i.e. equal to about £200,000 of our money] from Alexander as his contribution towards perfecting his *History of Animals*'.[15] Aelian, on the other hand, who lived at a period a little anterior to Athenaeus, tells us that it was 'Philip of Macedon who so esteemed learning that he supplied Aristotle with ample funds' adding that he similarly honoured both Plato and Theophrastus.[16]

Now in all Aristotle's works there is not a single sentence in praise of Alexander and there is some evidence that the two had become estranged. In support of this we may quote Plutarch (*c.* A. D. 100) who gives a detailed description of a conspiracy in 327 B. C. against Alexander by Callisthenes, a pupil of Aristotle who appears to have kept up a correspondence with his master.[17] Alexander himself wrote of Callisthenes, according to Plutarch: 'I will punish this sophist, together with those who sent him to me and those who harbour in their cities men who conspire against my life' and Plutarch adds that Alexander 'directly reveals in these words a hostility to Aristotle in whose house Callisthenes ... had been reared, being a son of Hero who was a niece of Aristotle'.[18] Yet the Alexandrian conquests, bringing Greece into closer contact with a wider world and extending Greek knowledge of the Orient, must have had their influence in stimulating interest in rare and curious creatures and in a general extension of natural knowledge. That the interest in these topics extended beyond the circle of the Peripatetics is shown by the fact that Speusippus, the pupil of Plato and his successor as leader of his school, occupied himself with natural history and wrote works on biological topics and especially on fish.

Nevertheless, remarkable as is Aristotle's acquaintance with animal forms, investigation shows that he is reliable only when treating of creatures native to the Aegean basin. As soon as he gets outside that area his statements are almost always founded on hearsay or even on fable.[19] Whatever assistance Aristotle may have received in the preparation of his biological works came, therefore, probably from no such picturesque and distant source as the gossip of Pliny or Aelian would suggest. We can conjecture that he received aid from the powerful relatives of his wife at Atarneus and in Lesbos, and we may most reasonably suppose that after his return to Athens much help would have been given him by his pupils within the Lyceum. To them may probably be ascribed many passages in the biological writings; for it seems hardly possible that Aristotle himself would have had time for detailed biological research after he had settled as a teacher in Athens. Of the work of these members of his school a fine monument has survived in two complete botanical treatises and fragments of others on zoological and psychological subjects by Theophrastus of Eresus, his pupil and successor in the leadership of the Lyceum and perhaps his literary legatee.

When we turn to the Aristotelian biological works themselves we naturally inquire first into the question of genuineness, and here a difficulty arises in that all his extant works have come down to us in a state that is not comparable to those of any other great writer. Among the ancients admiration was expressed for Aristotle's eloquence and literary powers, but, in the material that we have here to consider, very little trace of these qualities can be detected by even the most lenient judge. The arrangement of the subject-matter is far from perfect even if we allow for the gaps and disturbances caused by their passage through many hands. Moreover, there is much repetition and often irrelevant digression, while the language is usually plain to baldness and very frequently obscure. We find sometimes the lightening touch of humour, but the style hardly ever rises to beauty. Furthermore, even in matters of fact, while many observations exhibit wonderful insight and, forestalling modern discovery, betray a most searching and careful application of scientific methods, yet elsewhere we find errors that are childish and could have been avoided by the merest tyro.

This curious state of the Aristotelian writings has given rise to much discussion among scholars and to explain it there has been developed what is known as the 'notebook theory'. It is supposed that the bases of the material that we possess were notebooks put together by Aristotle himself for his own use, probably while lecturing. These passed, it is believed, into the hands of certain of his pupils and were perhaps in places incomprehensible as they stood. Such pupils, after the master's death, filled out the notebooks either from the memory of his teaching or from their own knowledge—or ignorance. Thus modified, however, they were still not prepared for publication, even in the limited sense in which works may be said to have been published in those days, but they formed again the fuller bases of notes for lectures delivered by his successors. In this form they have finally survived to our time, suffering, however, from certain further losses and displacements on a larger scale. Some of the 'Aristotelian' works are undoubtedly more deeply spurious, but the works that are regarded as 'genuine' do not seem to have been seriously tampered with, except by mere scribal or bookbinders' blunders, at any date later than a generation or two following Aristotle's own time. These notebooks as they stand are in fact probably in much the state in which we should find them were we able to retrieve a copy dating from the first or second century B. C.[20]

In the opening chapter of one of his great biological works Aristotle sets forth in detail his motives for the study of living things. The passage is in itself noteworthy as one of the few instances in which he rises to real eloquence.

'Of things constituted by nature some are ungenerated, imperishable, and eternal, while others are subject to generation and decay. The former are excellent beyond compare and divine, but less accessible to knowledge. The evidence that might throw light on them, and on the problems which we long to solve respecting them, is furnished but scantily by sensation; whereas respecting perishable plants and animals we have abundant information, living as we do in their midst, and ample data may be collected concerning all their various kinds, if only we are willing to take sufficient pains. Both departments, however, have their special charm. The scanty conceptions to which we can attain of celestial things give us, from their excellence, more pleasure than all our knowledge of the world in which we live; just as a half glimpse of persons we love is more delightful than a leisurely view of other things, whatever their number and dimensions. On the other hand, in certitude and in completeness our knowledge of terrestrial things has the advantage. Moreover, their greater nearness and affinity to us balances somewhat the loftier interest of the heavenly things that are the objects of the higher philosophy.... For if some [creatures] have no graces to charm the sense, yet even these, by disclosing to intellectual perception the artistic spirit that designed them, give immense pleasure to all who can trace links of causation, and are inclined to philosophy. We therefore must not recoil with childish aversion from the examination of the humbler animals. Every realm of nature is marvellous. It is told of Heraclitus that when strangers found him warming himself at the kitchen fire and hesitated to go in, he bade them enter since even in the kitchen divinities were present. So should we venture on the study of every kind of animal without distaste, for each and all will reveal to us something natural and something beautiful.[21] Absence of haphazard and conduciveness of everything to an end are to be found in Nature's works in the highest degree, and the resultant end of her generations and combinations is a form of the beautiful.

'If any person thinks the examination of the rest of the animal kingdom an unworthy task, he must hold in like disesteem the study of man. For no one can look at the primordia of the human frame—blood, flesh, bones, vessels, and the like—without much repugnance. Moreover, when any one of the parts or structures, be it which it may, is under discussion, it must not be supposed that it is its material composition to which attention is being directed or which is the object of the discussion, but the relation of such part to the total form....

'As every instrument and every bodily member subserves some partial end, that is to say, some special action, so the whole body must be destined to minister to some plenary sphere of action. Thus the saw is made for sawing, since sawing is a function, and not sawing for the saw. Similarly, the body too must somehow or other be made for the soul, and each part of it for some subordinate function to which it is adapted.'[22]

Aristotle is, in the fullest sense a 'vitalist'. He believes that the presence of a certain peculiar principle of a non-material character is essential for the exhibition of any of the phenomena of life. This principle we may call *soul*, translating his word ψυχή. Living things, like all else in nature, have, according to Aristotle, an end or object. 'Everything that Nature makes,' he says, 'is means to an end. For just as human creations are the products of art, so living objects are manifestly the products of an analogous cause or principle.... And that the heaven, if it had an origin, was evolved and is maintained by such a cause, there is, therefore, even more reason to believe, than that mortal animals so originated. For order and definiteness are much more manifest in the celestial bodies than in our own frame.'[23] It was a misinterpretation of this view that especially endeared him to the mediaeval Church and made it possible to absorb Aristotelian philosophy into Christian theology. It must be remembered that the cause or principle that leads to the development of living things is in Aristotle's view, not external but *internal*.

While putting his own view Aristotle does not fail to tell us of the standpoint of his opponents. 'Why, however, it must be asked, should we look on the operations of Nature as dictated by a final cause, and intended to realize some desirable end? Why may they not be merely the results of necessity, just as the rain falls of necessity, and not that the corn may grow? For though the rain makes the corn grow, it no more occurs in order to cause that growth, than a shower which spoils the farmer's crop at harvest-time occurs in order to do that mischief. Now, why may not this, which is true of the rain, be true also of the parts of the body? Why, for instance, may not the teeth grow to be such as they are merely of necessity, and the fitness of the front ones with their sharp edge for the comminution of the food, and of the hind ones with their flat surface for its mastication, be no more than an accidental coincidence, and not the cause that has determined their development?'[24]

The answers to these questions form a considerable part of Aristotle's philosophy where we are unable to follow him. For the limited field of biology, however, the question is on somewhat narrower lines. 'What,' he asks, 'are the forces by which the hand or the body was fashioned into shape? The wood carver will perhaps say, by the axe or the auger.... But it is not enough for him to say that by the stroke of his tool this part was formed into a concavity, that into a flat surface; but he must state the reasons why he struck his blow in such a way as to effect this and what his final object was ... [similarly] the true method [of biological science] is to state what the definite characters are that distinguish the animal as a whole; to explain what it is both in substance and in form, and to deal after the same fashion with its several organs.... If now this something, that constitutes the form of the living being, be the soul, or part of the soul, or something that, without the soul, cannot exist, (as would seem to be the case, seeing at any rate that when the soul departs, what is left is no longer a living animal, and that none of the parts remain what they were before, excepting in mere configuration, like the animals that in the fable are turned into stone;) ... then it will come within the province of the natural philosopher to inform himself concerning the soul, and to treat of it, either in its entirety, or, at any rate, of that part of it which constitutes the essential character of an animal; and it will be his duty to say what this soul or this part of a soul is.'[25] Thus in the Aristotelian writings the discussion of the nature and orders of 'soul' is almost inseparable from the subjects now included under the term Biology.

There can be no doubt that through much of the Aristotelian writings runs a belief in a *kinetic* as distinct from a static view of existence. It cannot be claimed that he regarded the different kinds of living things as actually passing one into another, but there can be no doubt that he fully realized that the different kinds can be arranged in a series in which the gradations are easy. His scheme would be something like that represented on p. 30 (Fig. 7 a).

'Nature,' he says, 'proceeds little by little from things lifeless to animal life in such a way that it is impossible to determine the exact line of demarcation, nor on which side thereof an intermediate form should lie. Thus, next after lifeless things in the upward scale comes the plant, and of plants one will differ from another as to its amount of apparent vitality; and, in a word, the whole *genus* of plants, whilst it is devoid of life as compared with an animal, is endowed with life as compared with other corporeal entities. Indeed, there is observed in plants a continuous scale of ascent towards the animal. So, in the sea, there are certain objects concerning which one would be at a loss to determine whether they be animal or vegetable.'[26]

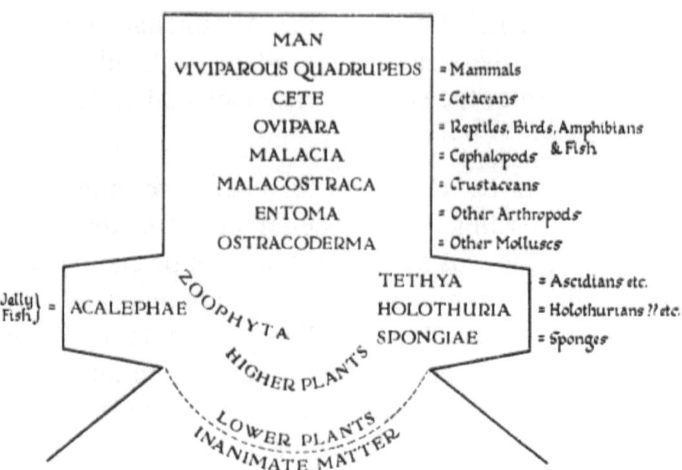

Fig. 7a. The Order of Living Things according to Aristotle.

'A sponge, in these respects completely resembles a plant, in that ... it is attached to a rock, and that when separated from this it dies. Slightly different from the sponges are the so-called Holothurias ... as also sundry other sea-animals that resemble them. For these are free and unattached, yet they have no feeling, and their life is simply that of a plant separated from the ground. For even among land-plants there are some that are independent of the soil— or even entirely free. Such, for example, is the plant which is found on Parnassus, and which some call the Epipetrum [probably *Sempervivum tectorum*, the common houseleek]. This you may hang up on a peg and it will yet live for a considerable time. Sometimes it is a matter of doubt whether a given organism should be classed with plants or with animals. The Tethya, for instance, and the like, so far resemble plants as that they never live free and unattached, but, on the other hand, inasmuch as they have a certain flesh-like substance, they must be supposed to possess some degree of sensibility.'[27]

'The Acalephae or Sea-nettles, ... lie outside the recognized groups. Their constitution, like that of the

Tethya, approximates them on the one side to plants, on the other side to animals. For seeing that some of them can detach themselves and can fasten on their food, and that they are sensible of objects which come in contact with them, they must be considered to have an animal nature.... On the other hand, they are closely allied to plants, firstly by the imperfection of their structures, secondly by their being able to attach themselves to the rocks, which they do with great rapidity, and lastly by their having no visible residuum notwithstanding that they possess a mouth.'[28]

Thus 'Nature passes from lifeless objects to animals in such unbroken sequence, interposing between them beings which live and yet are not animals, that scarcely any difference seems to exist between two neighbouring groups owing to their close proximity.'[29]

Some approach to evolutionary doctrine is also foreshadowed by Aristotle in his theories of the development of the individual. This is obscured, however, by his peculiar view of the nature of procreation. On this topic his general conclusion is that the material substance of the embryo is contributed by the female, but that this is mere passive formable material, almost as though it were the soil in which the embryo grows. The male by giving the principle of life, the soul, contributes the essential generative agency. But this *soul* is not material and it is, therefore, not theoretically necessary for anything material to pass from male to female. The material which does in fact so pass with the seed of the male is an accident, not an essential, for the essential contribution of the male is not matter but *form* and *principle*. The female provides the *material*, the male the *soul*, the *form*, the *principle*, that which makes life. Aristotle was thus prepared to accept instances of fertilization without material contact.

'The female does not contribute semen to generation but does contribute something ... for there must needs be that which generates and that from which it generates.... If, then, the male stands for the effective and active, and the female, considered as female, for the passive, it follows that what the female would contribute to the semen of the male would not be semen but material for the semen to work upon....

'How is it that the male contributes to generation, and how is it that the semen from the male is the cause of the offspring? Does [the semen] exist in the body of the embryo as a part of it from the first, mingling with the material which comes from the female? Or does the semen contribute nothing to the material body of the embryo but only to the power and movement in it?... The latter alternative appears to be the right one both *a priori* and in view of the facts.'[30]

This discussion leads to the question of the natural process of generation itself. It is a topic that we have seen discussed by an earlier writer who had set forth a sort of doctrine of pangenesis (see p. 14). His view Aristotle declines to share. 'We must', he says, 'say the opposite of what the ancients said. For whereas they said that semen is that which *comes from all* the body, we shall say that it is that whose nature is to *go to all* of it, and what they thought a waste-product seems rather to be a secretion.' According to Aristotle semen is derived from the same nutritive material in the blood vessels that is distributed to the rest of the body. The semen, however, is strained or secreted off from this nutritive material—as being its most essential and representative portion—before the distribution actually takes place.[31] But why, it may be asked, if the semen does not come from the various parts of the body, is it yet able to reproduce those various parts? The answer, on the Aristotelian view, seems to be that the semen contains special and peculiar fractions of the nutritive fluid which have been so modified and adapted that, if not secreted off as semen, they would be distributed to the different parts of the body to nourish each of these various parts. These substances have been elaborated by the *soul* or vital principle in a manner that is specifically suited for each organ, hand, liver, face, heart, &c., and from each of these specific substances a specific essence is separated off into the semen corresponding to hand, liver, face, heart, &c., of the offspring.

The next question that arises is the mechanism by which the offspring come to resemble their parents. The mechanism in the case of inheritance from the father is comprehensible when we consider the origin and nature of the semen, but the inheritance from the mother requires further explanation. The view of Aristotle is based upon the nature of the catamenia and their disappearance during gestation. 'The catamenia', in his view, 'are a secretion as the semen is.'[32] The female contributes the material by which the embryo grows and she does this through the catamenia which are suspended during gestation for this very purpose. The matter is thus summed up by Aristotle.

'The male does not emit semen at all in some animals, and where he does, this is no part of the resulting embryo; just so no material part comes from the carpenter to the material, i.e. to the wood in which he works, nor does any part of the carpenter's art exist within what he makes, but the shape and the form are imparted from him to the material by means of the motion he sets up. It is his hands that move his tools, his tools that move the material; it is his knowledge of his art, and his *soul*, in which is the form, that move his hands or any other part of him with a motion of some definite kind, a motion varying with the varying nature of the object made. In like manner, in the male of those animals which emit semen, Nature uses the semen as a tool and as possessing motion in actuality, just as tools are used in the products of any art, for in them lies in a certain sense the motion of the art.'[33]

'For the same reason the development of the embryo takes place in the female; neither the male himself nor the female emits semen into the female, but the female receives within herself the share contributed by both, because in the female is the material from which is made the resulting product. Not only must the mass of material from which the embryo is in the first instance formed exist there, but further material must constantly be added so that the embryo may increase in size. Therefore the birth must take place in the female. For the carpenter must keep in close connexion with his timber and the potter with his clay, and generally all workmanship and the ultimate movement imparted to matter must be connected with the material concerned, as, for instance, architecture is *in* the buildings it makes.'[34]

The problem of the nature of generation is one in which Aristotle never ceased to take an interest, and among the methods by which he sought to solve it was embryological investigation. In his ideas on the methods of reproduction we must seek also the main bases of such classification of animals as he exhibits. His most important embryological researches were made upon the chick. He asserts that the first signs of development are noticeable on the third day, the heart being visible as a palpitating blood-spot whence, as it develops, two meandering blood vessels extend to the surrounding tunics.

'Generation from the egg', he says, 'proceeds in an identical manner with all birds.... With the common hen after three days and nights there is the first indication of the embryo.... The heart appears like a speck of blood in the white of the egg. This point beats and moves as though endowed with life, and from it two vessels with blood in them trend in a convoluted course ... and a membrane carrying bloody fibres now envelops the yolk, leading off from the vessels.'[35]

Aristotle lays considerable stress on the early appearance of the heart in the embryo. Corresponding to the general gradational view that he had formed of Nature, he held that the most primitive and fundamentally important organs make their appearance before the others. Among the organs all give place to the heart, which he considered 'the first to live and the last to die'.[36]

A little later he observed that the body had become distinguishable, and was at first very small and white.

'The head is clearly distinguished and in it the eyes, swollen out to a great extent.... At the outset the under portion of the body appears insignificant in comparison with the upper portion....

'When an egg is ten days old the chick and all its parts are distinctly visible. The head still is larger than the rest of the body and the eyes larger than the head. At this time also the larger internal organs are visible, as also the stomach and the arrangement of the viscera; and the vessels that seem to proceed from the heart are now close to the navel. From the navel there stretch a pair of vessels, one [vitelline vein] towards the membrane that envelops the yolk, and the other [allantoic vein] towards that membrane which envelops collectively the membrane wherein the chick lies, the membrane of the yolk and the intervening liquid.... About the twentieth day, if you open the egg and touch the chick, it moves inside and chirps; and it is already coming to be covered with down when, after the twentieth day, the chick begins to break the shell.'[37]

Aristotle recognized a distinction in the mode of development of mammals from that of all other viviparous creatures. Having divided the apparently viviparous animals into two groups, one of which is truly and internally and the other only externally viviparous, he pointed out that in the mammalia, the group regarded by him as internally viviparous, the foetus is connected until birth with the wall of the mother's womb by the navel-string. These animals, in his view, produce their young without the intervention of an ovum, the embryo being 'living from the first'. Such non-mammals, on the other hand, as are viviparous are so in the external sense only, that is, the young which he considered to arise in this group from ova may indeed develop within the mother's womb and be born alive, but they go through their development without organic connexion with the mother's body, so that her womb acts but as a nursery or incubator for her eggs. It was indeed a sort of accident among the ovipara whether in any particular species the ovum went through its development inside or outside the mother's body. 'Some of the ovipara', he says, 'produce the egg in a perfect, others in an imperfect state, but it is perfected outside the body as has been stated of fish.'[38]

Yet though Aristotle regarded fish as an oviparous group, he knew also of kinds of fish that were externally viviparous. It is most interesting to observe, moreover, that he was acquainted with one particular instance among fish in which matters were less simple and in which the development bore an analogy to that of the mammalia, his true internal vivipara. 'Some animals', he says, 'are viviparous, others oviparous, others vermiparous. Some are viviparous, such as man, the horse, the seal and all other animals that are hair-coated, and, of marine animals, the Cetaceans, as the dolphin, *and the so-called Selachia.*'[39]

Aristotle tells us elsewhere that a species of these Selachia which he calls *galeos*—a name still used for the dog-fish by Greek fishermen—'has its eggs in betwixt the [two horns of the] womb; these eggs shift into each of the two horns of the womb and descend, and the young develop with the navel-string attached to the womb, so that, as the egg-substance gets used up, the embryo is sustained to all appearances just as in quadrupeds. The navel-string is ...

attached as it were by a sucker, and also to the centre of the embryo in the place where the liver is situated.... Each embryo, as in the case of quadrupeds, is provided with a chorion and separate membranes.'[40]

The remarkable anatomical relationship of the embryo of *Galeus* (*Mustelus*) *laevis* to its mother's womb was little noticed by naturalists until the whole matter was taken up by Johannes Müller about 1840.[41] That great observer demonstrated the complete accuracy of Aristotle's description and the justice of his comparison to and contrast with the mammalian mode of development.[42] The work of Johannes Müller at once had the effect of drawing the attention of naturalists to the importance and value of the Aristotelian biological observations.

Aristotle attempts to explain the viviparous character of the Selachians. His explanation has perhaps little meaning for the modern biologist, just as many of our scientific explanations will seem meaningless to our successors. But such explanations are often worth consideration not only as stages in the historical development of scientific thought, but also as illustrating the fact that while the ultimate object of science is a *description* of nature, the immediate motive of the best scientific work is usually an *explanation* of nature. Yet it is usually the descriptive, not the explanatory element that bears the test of time.

'Birds and scaly reptiles', says Aristotle, 'because of their heat produce a perfect egg, but because of their dryness it is only an egg. The cartilaginous fishes have less heat than these but more moisture, so that they are intermediate, for they are both oviparous and viviparous within themselves, the former because they are cold, the latter because of their moisture; for moisture is vivifying, whereas dryness is farthest removed from what has life. Since they have neither feathers nor scales such as either reptiles or other fishes have, all of which are signs rather of a dry and earthy nature, the egg they produce is soft; for the earthy matter does not come to the surface in their eggs any more than in themselves. That is why they lay eggs in themselves, for if the egg were laid externally it would be destroyed, having no protection.'[43]

This explanation is based on Aristotle's fundamental doctrine of the opposite *qualities*, heat, cold, wetness, and dryness, that are found combined in pairs in the four *elements*, earth, air, fire, and water. The theory was of the utmost importance for the whole subsequent development of science and was not displaced until quite modern times. It was not an original conception of Aristotle, for something resembling it had been set forth long before his time in figurative language by Empedocles (*c.* 500-*c.* 430 B. C.), as Aristotle himself tells us.[44] The same view had been foreshadowed by Pythagoras (*c.* 580-*c.* 490 B. C.) at an even earlier date and was perhaps of much greater antiquity. But Aristotle developed the doctrine and was the main channel for its conveyance to later ages, so that his name will always be associated with it. Matter in general and living matter in particular was held by him to be composed of these four essential so-called *elements* (στοιχεῖ), each of which is in turn compounded from two of the primary *qualities* (δυνάμεις) which Aristotle brought into relation with the elements. Thus earth was cold and dry, water cold and wet, air hot and wet, and fire hot and dry (<u>Fig. 7b</u>).

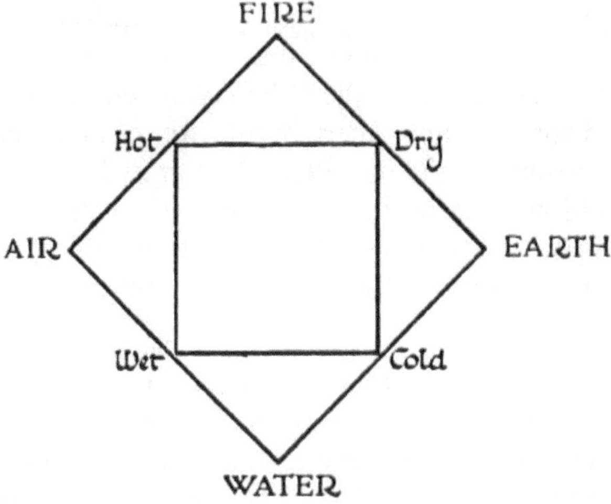

Fig. 7b. The Four Elements and the Four Qualities.

The theory of the elements and qualities is applicable to all matter and not specially to living things. The distinction between the living and not-living is to be sought not so much in its material constitution, but in the presence or absence of 'soul', and his teaching on that topic is to be found in his great work περὶ ψυχῆς, *On Soul*. He does not think of matter as organic or inorganic—that is a distinction of the seventeenth century physiologists—nor does he think of things as divided into animal, vegetable, and mineral—that is a distinction of the mediaeval alchemists,—but he thinks of things as either with soul or without soul (ἔμψυχα or ἄαψυχα).

His belief as to the relationship of this soul to material things is a difficult and complicated subject which would take us far beyond the topics included in biological writings to-day, but he tells us that 'there is a class of existent things which we call substance, including under that term, firstly, matter, which in itself is not this nor that; secondly, shape or form, in virtue of which the term this or that is at once applied; thirdly, the whole made up of matter and form. Matter is identical with potentiality, form with actuality,' the soul being, in living things, that which gives the form or actuality. 'Of natural bodies', he continues, 'some possess life and some do not: where by life we mean the power of self-nourishment and of independent growth and decay'.[45] It should here be noted that in the Aristotelian sense the ovum is not at first a living thing, for in its earliest stage and before fertilization it does not possess soul even in its most elementary form.

'The term life is used in various senses, and, if life is present in but a single one of these senses, we speak of a thing as living. Thus there is intellect, sensation, motion from place to place and rest, the motion concerned with nutrition, and, further, [there are the processes of] decay and growth,' all various meanings or at least exhibitions of some form of life. Hence even 'plants are supposed to have life, for they have within themselves a faculty and principle whereby they grow and decay.... They grow and continue to live so long as they are capable of absorbing nutriment. This form of life can be separated from the others ... and plants have no other faculty of soul at all,' but only this lowest vegetative soul. 'It is then in virtue of this principle that all living things live, whether animals or plants. But it is sensation which primarily constitutes the animal. For, provided they have sensation, even those creatures that are devoid of movement and do not change their place are called animals.... As the nutritive faculty may exist without touch or any form of sensation, so also touch may exist apart from other senses.'[46] Apart from these two lower forms of soul, the *vegetative* or nutritive and reproductive and the *animal* or sensitive, stands the *rational* or intellectual soul peculiar to man, a form of soul with which we shall here hardly concern ourselves.[47]

The possession of one or more of the three types of soul, vegetative, sensitive, and rational, provides in itself a basis for an elementary form of arrangement of living things in an ascending scale. We have already seen that Aristotle certainly describes something resembling a 'Scala Naturae' and that such a scheme can easily be drawn up from passages in his works. It may, however, be doubted whether his phraseology is capable of extension so as to include a true *classification* of animals in any modern sense. It is true that he repeatedly divides animals into classes, *Sanguineous* and *Non-sanguineous*, *Oviparous* and *Viviparous*, *Terrestrial* and *Aquatic*, &c., but his divisions are for the most part simply dichotomic. He certainly defines a few groups of animals as the Lophura (*Equidae*), the Cete (*Cetacea*), and the Selache (*Elasmobranchiae* together with the *Lophiidae*) in a way that fairly corresponds to similar groups in later systems. In most cases, however, his definitions are not exact enough for modern needs, for the same animal may fall into more than one of his classes and widely different animals into the same class. Thus he invents a category *Carcharodonta* for animals with sharp interlocking teeth and includes in it carnivores, reptiles, and fish; again, the horse kind must be included both among his *Anepallacta* or animals having flat crowned teeth as well as among the *Amphodonta* or animals with front teeth in both jaws. Such words as these are really terms of *description*, not of classification in the modern biological sense of that word.

There are, however, scattered through the biological works, certain terms which are applied to animal groups and organs and are defined in such a way as to suggest that they might ultimately have been developed for classificatory purposes. Thus his lowest group is the *species*. 'The individuals comprised within a single *species* (εἶδος) ... are the real existences; but inasmuch as these individuals possess one common specific form, it will suffice to state the universal attributes of the species, that is, the attributes common to all its individuals, once and for all.'[48] This is surely not very far removed from the modern biological conception of a species.

'But as regards the larger groups—such as birds—which comprehend many species, there may be a question. For on the one hand it may be urged that as the ultimate species represent the real existences, it will be well, if practicable, to examine these ultimate species separately, just as we examine the species Man separately; to examine, that is, not the whole class Birds collectively, but the Ostrich, the Crane, and the other indivisible groups or species belonging to the class.

'On the other hand, this course would involve repeated mention of the same attribute, as the same attribute is common to many species, and so far would be somewhat irrational and tedious. Perhaps, then, it will be best to treat generically the universal attributes of the groups that have a common nature and contain closely allied subordinate forms, whether they are groups recognized by a true instinct of mankind, such as Birds and Fishes, or groups not popularly known by a common appellation, but withal composed of closely allied subordinate groups; and only to deal individually with the attributes of a single species, when such species—man, for instance, and any other such, if such there be—stands apart from others, and does not constitute with them a larger natural group.

'It is generally similarity in the shape of particular organs, or of the whole body, that has determined the formation of the larger groups. It is in virtue of such a similarity that Birds, Fishes, Cephalopoda, and Testacea have been made to form each a separate *genus* (γένος). For within the limits of each such *genus*, the parts do not differ in that they have no nearer resemblance than that of analogy—such as exists between the bone of man and the spine of fish—but they differ merely in respect of such corporeal conditions as largeness smallness, softness hardness, smoothness roughness, and other similar oppositions, or, in one word, in respect of degree.'[49]

The Aristotelian *genus* thus differs widely from the term as used in modern biology. In another passage he comes nearer to defining it and the analogy of parts which extends from genus to genus.

'Groups that differ only in the degree, and in the more or less of an identical element that they possess are aggregated together under a single *genus*; groups whose attributes are not identical but *analogous* are separated. For instance, bird differs from bird by gradation, or by excess and defect; some birds have long feathers, others short ones, but all are feathered. Bird and Fish are more remote and only agree in having analogous organs; for what in the bird is feather, in the fish is scale. Such *analogies* can scarcely, however, serve universally as indications for the formation of groups, for almost all animals present analogies in their corresponding parts.'[50]

Aristotle nowhere gives to his term *genus* a rigid application that can be applied throughout the animal kingdom. He uses the word in fact much as we should use the conveniently flexible term *group*, now for a larger and less definite, now for a smaller and more definite collection of species. This varying use of a technical word makes it impossible to draw up a classification based on his *genera* or indeed with any consistent use of the terms which he actually employs.

The difficulty or impossibility of drawing up a satisfactory classificatory system from the Aristotelian writings has not, however, deterred numerous naturalists and scholars from making the attempt, and the subject has in itself a considerable history and literature[51] extending from the days of Edward Wotton (1492-1555) downward.[52] The more recent efforts at drawing up an Aristotelian classificatory system have been based on the methods of reproduction to which he certainly attached very great importance.[53] Provided that it be remembered that Aristotle does not himself detail any such system there can be no harm in constructing one from his works. At worst it will serve as a *memoria technica* for the extent and character of his knowledge of natural history, and at best it may represent a scheme to which he was tending.

ENAIMA (*Sanguineous and either viviparous oroviparous*)
    = *vertebrates.*

1.ἄνθρωπος. Man.
2.κήτη. Cetaceans.
3. ζῷα τετράποδα ζωοτόκα ἐν αὑτοῖς.
Viviparous quadrupeds.
 (a) μὴ ἀμφώδοντα. Non-amphodonts = Ruminants with incisor in lower jaw only and with cloven hoofs.
 (b) μώνυχα. Solid-hoofed animals.
  i. λόφουρα. Equidae.
  ii. μώνυχα ἕτερα. Other solid-hoofed animals.
4.ὄρνιθες. Birds.

Viviparous in the internal sense.

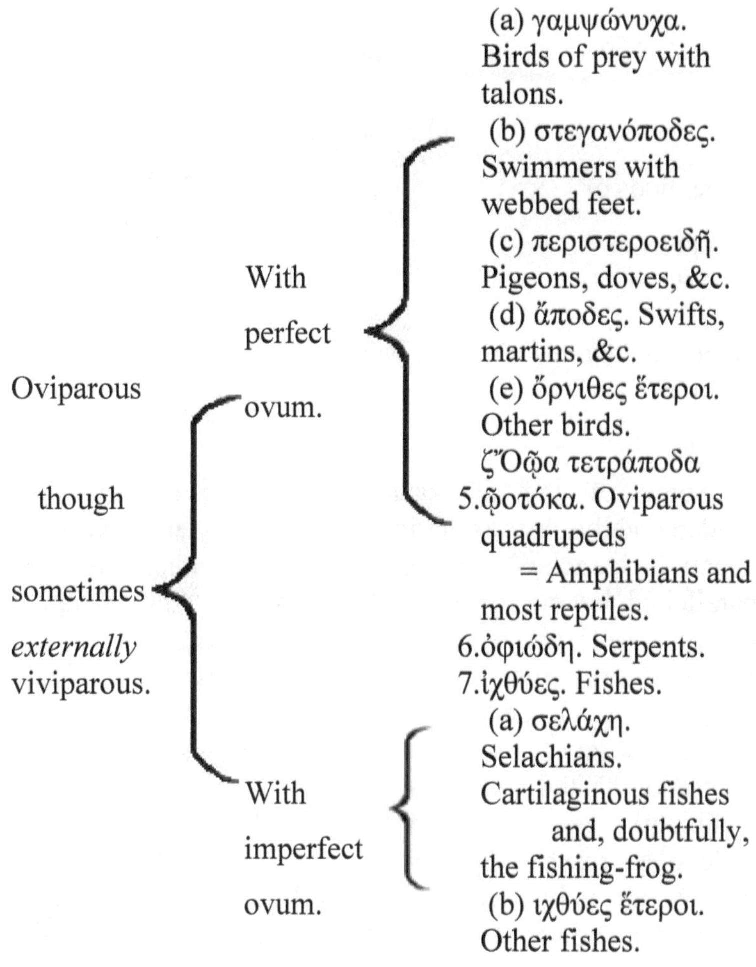

ANAIMA (Non-sanguineous and either viviparous, vermiparous or budding)
= *Invertebrates*

With perfect ovum.   { 8. μαλάκια. Cephalopods.
               { 9. μαλακόστρακα. Crustaceans.

With 'scolex'.	10.	ἔντομα. Insects, spiders, scorpions, &c.
With generative slime, buds or spontaneous generation.	11.	ὀστρακόδερμα. Molluscs (except Cephalopods), Echinoderms, &c.
With spontaneous generation only.	12.	ζωόφυτα. Sponges, Coelenterates, &c.

Some of the elements in this classification are fundamentally unsatisfactory in that they are based on negative characters. Such is the group of *Anaima* which is parallelled by our own equally convenient and negative though morphologically meaningless equivalent *Invertebrata*. Others, such as the subdivisions of the viviparous quadrupeds, can only be forcibly extracted out of Aristotle's text. But there are yet others, such as the separation of the cartilaginous from the bony fishes, that exhibit true genius and betray a knowledge that can only have been reached by careful investigation. Remarkably brilliant too is his treatment of Molluscs. There can be no doubt that he dissected the bodies and carefully watched the habits of octopuses and squids, *Malacia* as he calls them. He separates them too far from the other Molluscs, grouped by him as *Ostracoderma*, but his actual descriptions of the structure and sexual process of the cephalopods are exceedingly remarkable, and after being long disregarded or misunderstood were verified and repeated in the course of the nineteenth century.[54]

Passing from his general ideas on the nature and division of living creatures we may turn to some of the most noteworthy of his actual observations. In the realm of comparative anatomy proper we may instance that of the stomach of ruminants. He must have dissected these animals, for he gives a clear and correct account of the four chambers. 'Animals', he says, 'present diversities in the structure of their stomachs. Of the viviparous quadrupeds, such of the horned animals as are not equally furnished with teeth in both jaws are furnished with four such chambers. These animals are those that are said to chew the cud. In these animals the oesophagus extends from the mouth downwards along the lung, from the midriff to the *big stomach* [*rumen*, or paunch], and this stomach is rough inside and semi-partitional. And connected with it near to the entry of the oesophagus is what is called the *kekryphalos* [*reticulum*, or honeycomb bag]; for outside it is like the stomach, but inside it resembles a netted cap; and the kekryphalos is a good deal smaller than the *big stomach*.' The term *kekryphalos* was applied to the net that women wore over their hair to keep it in order. 'Connected with this kekryphalos,' he continues, 'is the *echinos* [*psalterium*, or *manyplies*], rough inside and laminated, and of about the same size as the kekryphalos. Next after this comes what is called the *enystron* [*abomasum*], larger and longer than the echinos, furnished inside with numerous folds or ridges, large and smooth. After all this comes the gut....'[55] 'All animals that have horns, the sheep for instance, the ox, the goat, the deer and the like, have these several stomachs.... The several cavities receive the food one from the other in succession: the first taking the unreduced substances, the second the same when somewhat reduced, the third when reduction is complete, and the fourth when the whole has become a smooth pulp....'[56] 'Such is the stomach of those quadrupeds that are horned and have an unsymmetrical dentition (μὴ ἀμφώδοντα); and these animals differ one from another in the shape and size of the parts, and in the fact of the oesophagus reaching the stomach central-wise in some cases and sideways in others. Animals that are furnished equally with teeth in both jaws (ἀμφώδοντα) have one stomach; as man, the pig, the dog, the bear, the lion, the wolf.'[57]

A very famous example in the Aristotelian works anticipating modern biological knowledge is afforded by his reference to the mode of reproduction of the cephalopods. 'The Malacia such as the octopus, the sepia, and the calamary, have sexual intercourse all in the same way; that is to say, they unite at the mouth by an interlacing of their tentacles. When, then, the octopus rests its so-called head against the ground and spreads abroad its tentacles, the other sex fits into the outspreading of these tentacles, and the two sexes then bring their suckers into mutual connexion. Some assert that the male has a kind of penis in one of his tentacles, the one in which are the largest suckers; and they further assert that the organ is tendinous in character growing attached right up to the middle of the tentacle, and that the latter enables it to enter the nostril or funnel of the female.'[58]

The reproductive processes of the Cephalopods were unknown to modern naturalists until the middle of the nineteenth century. Before that time several observers had noted the occasional presence of a peculiar parasite in the mantle cavity of female cephalopods and had described its supposed structure without tracing any relationship to the process of generation. In 1851 it was first shown that this supposed parasite was the arm of the male animal specially modified for reproductive purposes and broken off on insertion into the mantle cavity of the female[59]. The actual process of reproduction does not seem to have been observed until 1894[60].

Aristotle is perhaps at his best and happiest when describing the habits of living animals that he has himself observed. Among his most pleasing accounts are those of the fishing-frog and torpedo. In these creatures he did not fail to notice the displacement of the fins associated with the depressed form of the body.

'In marine creatures,' he says, 'one may observe many ingenious devices adapted to the circumstances of their lives. For the account commonly given of the frog-fish or angler is quite true; as is also that of the torpedo....

'In the Torpedo and the Fishing-frog the breadth of the anterior part of the body is not so great as to render locomotion by fins impossible, but in consequence of it the upper pair [*pectorals*] are placed further back and the under pair [*ventrals*] are placed close to the head, while to compensate for this advancement they are reduced in size so as to be smaller than the upper ones.

'In the Torpedo the two upper fins [pectorals] are placed in the tail, and the fish uses the broad expansion of its body to supply their place, each lateral half of its circumference serving the office of a fin.... The torpedo narcotizes the creatures that it wants to catch, overpowering them by the force of shock that is resident in its body, and feeds upon them; it also hides in the sand and mud, and catches all the creatures that swim in its way and come under its narcotizing influence. This phenomenon has been actually observed in operation.... The torpedo-fish is known to cause a numbness even in human beings.

'The frog-fish has a set of filaments that project in front of its eyes; they are long and thin, like hairs, and are round at the tips; they lie on either side, and are used as baits.... The little creatures on which this fish feeds swim up to the filaments, taking them for bits of seaweed such as they feed upon. Accordingly, when the frog-fish stirs himself up a place where there is plenty of sand and mud and conceals himself therein, it raises the filaments, and when the little fish strike against them the frog-fish draws them in underneath into its mouth.... That the creatures get their living by this means is obvious from the fact that, whereas they are peculiarly inactive, they are often caught with mullets, the swiftest of fishes, in their interior. Furthermore, the frog-fish is usually thin when he is caught after losing the tips of his filaments.'[61]

The modification of the musculature of the torpedo-fish for electric purposes and the fishing habits of the fishing-frog or *Lophius* are now well known, but it was many centuries before naturalists had confirmed the observations of the father of biology.

When we turn from Aristotle's observations in the department of natural history to his discussion of the actual mechanism of the living body, the subject now contained under the heading *Experimental Physiology*, we are in the presence of much less satisfactory material. Aristotle here exhibits his weakness in physics and not being endowed with any experimental knowledge of that subject his physiological development is very greatly handicapped. He seems often to accept fancies of his own in place of generalizations from collated observations. This tendency of his was conveyed to his successors and delayed physiological advance for many centuries. It forms a striking contrast to the method of certain of the Hippocratic works such as the *Epidemics* and the *Aphorisms* which exhibit an investigator intent on recording actual observations and on deducing general laws therefrom. Had the Hippocratic method been extended by Aristotle beyond the field of natural history, where he freely follows it, to that of physiology, the succeeding generations might have established medicine far more firmly as a science.

An important factor in Aristotle's physical and physiological teaching is the doctrine that matter is continuous and not made up of indivisible parts. He thus rejected the atomic views of his predecessors Leucippus and Democritus which have been preserved for us by the poem of Lucretius. The different kinds of matter existing merely in their state of simple mixture formed various uniform or homogeneous substances, *homoeomeria*, of which the *tissues* of living bodies provided one type. We now consider tissues as having structure made up of living cells or their products, but to Aristotle their structure was an essential fact following on their particular elemental constitution. The structure of muscle or flesh was perhaps comparable to that of a crystalline substance, for, as we have seen, Aristotle made no fundamental distinction between organic and inorganic *substances*, which are in his view alike subject to the processes of generation and corruption. The difference between them lies not in their structure but in their potential relation to the various degrees of soul, the vegetative, the animal, and the rational.

'There are', says Aristotle, 'three degrees of composition, and of these the first in order is composition out of what some call the *elements*, earth, air, water, and fire....

'The second degree of composition is that by which the *homogeneous* parts of animals (ὁμοιομερῆ), such as bone, flesh, and the like, are constituted out of [these] primary substances.

'The third and last stage is the composition which forms the *heterogeneous* parts (ἀνομοιομερῆ) such as face, hand, and the rest.'[62]

The distinctions are not altogether clear but may perhaps be explained along such lines as the following. The division into homogeneous and heterogeneous corresponds in a general way to the later division into Tissues and Organs, the former, however, including much that we should not call tissue. The homogeneous parts were again of two kinds: (*a*) simple tissues or stuffs without any notion of size or shape, that is, mere substance capable of endowment with life or soul, e.g. cartilaginous or osseous tissues; and (*b*) simple structure, that is actual structure made of such a single tissue but with definite form and size, matter to which form had been added and which either was actually or had been endowed with soul, e.g. *a* cartilage or *a* bone.

As a physiologist Aristotle is, in fact, in much the same position as he is as a physicist. He never dissected the human body, he had only the roughest idea of the course of the vessels, and his description of the vascular system is so difficult and confused that a considerable literature has been written on its interpretation. He regarded the heart as the central organ of the body and the seat of sensation and he probably believed that the arteries contained air as well as blood. He made no adequate distinction between veins and arteries. He tells us that two great vessels arise from the heart and that the heart is, as it were, a part of these vessels. The two vessels are apparently the aorta and the vena cava, and a very elementary and not very accurate description is given of the branches of these vessels. He believed that the heart had three chambers or cavities and that it took in air direct from the lung.

The brain was for him mainly an organ by which were secreted certain cold humours which prevented any overheating of the body by the furnace of the heart under the action of the bellows of the lung. He formally rejected the older views of Diogenes of Apollonia, of Alcmaeon of Croton, and of the Hippocratic writings, that placed the seat of sensation in the brain.[63] He failed to trace any adequate relation of sense organs and nerves to brain. He considered that the spinal marrow served to hold the vertebrae together.

In general we may say that his physiology is on a much lower plane than his natural history, since in dealing with physiological questions he always seems to have in mind the body as a whole and seldom pauses for any detailed investigation of a particular part. The physiological views of

Aristotle were far from being fully accepted even by the generation which followed him. There was already growing up a school of physiologists whose work culminated five centuries later in that of Galen, where we find quite other views of the bodily functions. It is these views which we may take as more typical of the bases of Greek physiology (<u>see p. 66</u>).

In much of the Aristotelian material that we have discussed we have seen the development of a class of interests very foreign to those of the modern biologist, in whose work the general discussion of the ultimate nature and origin of life seldom plays a large part. The business of the modern biologist is mainly with vital phenomena as he encounters them and he is not concerned with the deeper philosophical problems. The man of science considers a part of the Universe where the philosopher makes it his business to regard the whole. With Aristotle this modern scientific process of taking a part of the sensible Universe, such as a particular group of animals or the particular action of a particular organ, and considering it in and by and for itself without reference to other things, had not yet fully emerged. Philosophy and science are still inextricably linked and there is no clear demarcation between them.

This is at least his theoretical view. But besides being a philosopher by choice he was a supreme naturalist by his natural endowments and he cannot suppress his love for nature and his capacity for observation. We see Aristotle the naturalist at his greatest as a direct observer or when reasoning directly about the observations that he has made. When he disregards his own observations and begins to erect theories on the observations or the views of others, he becomes weaker and less comprehensible.

All Aristotle's surviving biological works refer primarily to the animal creation. His work on plants is lost or rather has survived as the merest corrupted fragment. We are fortunate, however, in the possession of a couple of complete works by his pupil and successor Theophrastus (372-287), which may not only be taken to represent the Aristotelian attitude towards the plant world, but also give us an inkling of the general state of biological science in the generation which succeeded the master.

These treatises of Theophrastus are in many respects the most complete and orderly of all ancient biological works that have reached our time. They give an idea of the kind of interest that the working scientist of that day could develop when inspired rather by the genius of a great teacher than by the power of his own thoughts. Theophrastus is a pedestrian where Aristotle is a creature of wings, he is in a relation to the master of the same order that the morphologists of the second half of the nineteenth century were to Darwin. For a couple of generations after the appearance of the *Origin of Species* in 1859 the industry and ability of naturalists all over the world were occupied in working out in detail the structure and mode of life of living things on the basis of the Evolutionary philosophy. Nearly all the work on morphology and much of that on physiology since his time might be treated as a commentary on the works of Darwin. These volumes of Theophrastus give the same impression. They represent the remains—alas, almost the only biological remains—of a school working under the impulse of a great idea and spurred by the memory of a great teacher. As such they afford a parallel to much scientific work of our own day, produced by men without genius save that provided by a vision and a hope and an ideal. Of such men it is impossible to write as of Aristotle. Their lives are summed up by their actual achievement, and since Theophrastus is an orderly writer whose works have descended to us in good state, he is a very suitable instance of the actual standard of achievement of ancient biology. 'Without vision the people perish' and the very breath of life of science is drawn, and can only be drawn, from that very small band of prophets who from time to time, during the ages, have provided the great generalizations and the great ideals. In this light let us examine the work of Theophrastus.

In the absence of any adequate system of classification, almost all botany until the seventeenth century consisted mainly of descriptions of species. To describe accurately a leaf or a root in the language in ordinary use would often take pages. Modern botanists have invented an elaborate terminology which, however hideous to eye and ear, has the crowning merit of helping to abbreviate scientific literature. Botanical writers previous to the seventeenth century were substantially without this special mode of expression. It is partly to this lack that we owe the persistent attempts throughout the centuries to represent plants pictorially in herbals, manuscript and printed, and thus the possibility of an adequate history of plant illustration.

Theophrastus seems to have felt acutely the need of botanical terms, and there are cases in which he seeks to give a special technical meaning to words in more or less current use. Among such words are *carpos* = fruit, *pericarpion* = seed vessel = pericarp, and *metra*, the word used by him for the central core of any stem whether formed of wood, pith, or other substance. It is from the usage of Theophrastus that the exact definition of fruit and pericarp has come down to us.[64] We may easily discern also the purpose for which he introduces into botany the term *metra*, a word meaning primarily the *womb*, and the vacancy in the Greek language which it was made to fill. '*Metra*,' he says, 'is that which is in the middle of the wood, being third in order from the bark and [thus] like to the marrow in bones. Some call it the *heart* (καρδίαν), others the *inside* (ἐντεριώνην), yet others call only the innermost part of the metra itself the heart, while others again call this *marrow*.'[65] He is thus inventing a word to cover all the different kinds of core and importing it from another study. This is the method of modern scientific nomenclature which hardly existed for botanists even as late as the sixteenth century of our era. The real foundations of our modern nomenclature were laid in the later sixteenth and in the seventeenth century by Cesalpino and Joachim Jung.

Theophrastus understood the value of developmental study, a conception derived from his master. 'A plant', he says, 'has power of germination in all its parts, for it has life in them all, wherefore we should regard them not for what they are but for what they are becoming.'[66] The various modes of plant reproduction are correctly distinguished in a way that passes beyond the only surviving earlier treatise that deals in detail with the subject, the Hippocratic work *On generation*. 'The manner of generation of trees and plants are these: spontaneous, from a seed, from a root, from a piece torn off, from a branch or twig, from the trunk itself, or from pieces of the wood cut up small.'[67] The marvel of generation must have awakened admiration from a very early date. We have already seen it occupying a more ancient author, and it had also been one of the chief pre-occupations of Aristotle. It is thus not remarkable that the process should impress Theophrastus, who has left on record his views on the formation of the plant from the seed.

'Some germinate, root and leaves, from the same point, some separately from either end of the seed. Thus wheat, barley, spelt, and all such cereals [germinate] from either end, corresponding to the position [of the seed] in the ear, the root from the stout lower part, the shoot from the upper; but the two, root and stem, form a single continuous whole. The bean and other leguminous plants are not so, but in them root and stem are from the same point, namely, their place of attachment to the pod, where, it is plain, they have their origin. In some cases there is a process, as in beans, chick peas, and especially lupines, from which the root grows downward, the leaf and stem upward.... In certain trees the bud first germinates within the seed, and, as it increases in size, the seeds split—all such seeds are, as it were, in two halves; again, all those of leguminous plants have plainly two lobes and are double—and then the root is immediately thrust out. But in cereals, the seeds being in one piece, this does not happen, but the root grows a little before [the shoot].

'Barley and wheat come up monophyllous, but peas, beans, and chick peas polyphyllous. All leguminous plants have a single woody root, from which grow slender side roots ... but wheat, barley, and the other cereals have numerous slender roots by which they are matted together.... There is a contrast between these two kinds; the leguminous plants have a single root and have many side-growths above from the [single] stem ... while the cereals have many roots and send up many shoots, but these have no side-shoots.'[68]

There can be no doubt that here is a piece of minute observation on the behaviour of germinating seeds. The distinction between dicotyledons and monocotyledons is accurately set forth, though the stress is laid not so much on the cotyledonous character of the seed as on the relation of root and shoot. In the dicotyledons root and shoot are represented as springing from the same point, and in monocotyledons from opposite poles in the seed.

No further effective work was done on the germinating seed until the invention of the microscope, and the appearance of the work of Highmore (1613-85),[69] and the much more searching investigations of Malpighi (1628-94)[70] and Grew (1641-1712)[71] after the middle of the seventeenth century. The observations of Theophrastus are, however, so accurate, so lucid, and so complete that they might well be used as legends for the plates of these writers two thousand years after him.

Much has been written as to the knowledge of the sex of plants among the ancients. It may be stated that of the sexual elements of the flower no ancient writer had any clear idea. Nevertheless, sex is often attributed to plants, and the simile of the Loves of Plants enters into works of the poets. Plants are frequently described as male and female in ancient biological writings also, and Pliny goes so far as to say that some students considered that all herbs and trees were sexual.[72] Yet when such passages can be tested it will be found that these so-called males and females are usually different species. In a few cases a sterile variety is described as the male and a fertile as the female. In a small residuum of cases diœcious plants or flowers are regarded as male and female, but with no real comprehension of the sexual nature of the flowers. There remain the palms, in which the knowledge of plant sex had advanced a trifle farther. 'With dates', says Theophrastus, 'the males should be brought to the females; for the males make the fruit persist and ripen, and this some call by analogy *to use the wild fig* (ὀλυνθάζειν).[73] The process is thus: when the male is in flower they at once cut off the spathe with the flower and shake the bloom, with its flower and dust, over the fruit of the female, and, if it is thus treated, it retains the fruit and does not shed it.'[74] The fertilizing character of the spathe of the male date palm was familiar in Babylon from a very early date. It is recorded by Herodotus[75] and is represented by a frequent symbol on the Assyrian monuments.

The comparison of the fertilization of the date palm to the use of the wild fig refers to the practice of Caprification. Theophrastus tells us that there are certain trees, the fig among them, which are apt to shed their fruit prematurely. To remedy this 'the device adopted is caprification. Gall-insects come out of the wild figs which are hanging there, eat the tops of the cultivated figs, and so make them swell'.[76] These gall-insects 'are engendered from the seeds'.[77] Theophrastus distinguished between the process as applied to the fig and the date, observing that 'in both [fig and date] the male aids the female—for they call the fruit-bearing [palm] *female*—but whilst in the one there is a union of the two sexes, in the other things are different'.[78]

Theophrastus was not very successful in distinguishing the nature of the primary elements of plants, though he was able to separate root, stem, leaf, stipule, and flower on morphological as well as to a limited extent on physiological grounds. For the root he adopts the familiar definition, the only one possible before the rise of chemistry, that it 'is that by which the plant draws up nourishment',[79] a description that applies to the account given by the pre-Aristotelian author of the work περὶ γονῆς, *On generation*. But Theophrastus shows by many examples that he is capable of following out morphological homologies. Thus he knows that the ivy regularly puts forth roots from the shoots between the leaves, by means of which it gets hold of trees and walls,[80] that the mistletoe will not sprout except on the bark of living trees into which it strikes its roots, and that the very peculiar formation of the mangrove tree is to be explained by the fact that 'this plant sends out roots from the shoots till it has hold on the ground and roots again: and so there comes to be a continuous circle of roots round the tree, not connected with the main stem, but at a distance from it'.[81] He does not succeed, however, in distinguishing the real nature of such structures as bulbs, rhizomes, and tubers, but regards them all as roots. Nor is he more successful in his discussion of the nature of stems. As to leaves, he is more definite and satisfactory, though wholly in the dark as to their function; he is quite clear that the pinnate leaf of the rowan tree, for instance, is a leaf and not a branch.

From VILLA ALBANI
Copy (second century A. D.?) of earlier work

Notwithstanding his lack of insight as to the nature of sex in flowers, he attains to an approximately correct idea of the relation of flower and fruit. Some plants, he says, 'have [the flower] around the fruit itself as vine and olive; [the flowers] of the latter, when they drop, look as though they had a hole through them, and this is taken for a sign that it has blossomed well; for if [the flower] is burnt up or sodden, the fruit falls with it, and so it does not become pierced. Most flowers have the fruit case in the middle, or it may be the flower is on the top of the pericarp as in pomegranate, apple, pear, plum, and myrtle ... for these have their seeds below the flower.... In some cases again the flower is on top of the seeds themselves as in ... all thistle-like plants'.[82] Thus Theophrastus has succeeded in distinguishing between the hypogynous, perigynous, and epigynous types of flower, and has almost come to regard its relation to the fruit as the essential floral element.

Theophrastus has a perfectly clear idea of plant distribution as dependent on soil and climate, and at times seems to be on the point of passing from a statement of climatic distribution into one of real geographical regions. The general question of plant distribution long remained at, if it did not recede from, the position where he left it. The usefulness of the manuscript and early printed herbals in the West was for centuries marred by the retention of plant descriptions prepared for the Greek East and Latin South, and these works were saved from complete ineffectiveness only by an occasional appeal to nature.

With the death of Theophrastus about 287 B. C. pure biological science substantially disappears from the Greek world, and we get the same type of deterioration that is later encountered in other scientific departments. Science ceases to have the motive of the desire to know, and becomes an applied study, subservient to the practical arts. It is an attitude from which in the end applied science itself must suffer also. Yet the centuries that follow were not without biological writers of very great ability. In the medical school of Alexandria anatomy and physiology became placed on a firm basis from about 300 B. C., but always in the position subordinate to medicine that they have since occupied. Two great names of that school, Herophilus and Erasistratus, we must consider elsewhere.[83] Their works have disappeared and we have the merest fragments of them. In the last pre-Christian and the first two post-Christian centuries, however, there were several writers, portions of whose works have survived and are of great biological importance. Among them we include Crateuas, a botanical writer and illustrator, who greatly developed, if he did not actually introduce, the method of representing plants systematically by illustration rather than by description. This method, important still, was even more important when there was no proper system of botanical nomenclature. Crateuas by his paintings of plants, copies of which have not improbably descended to our time, began a tradition which, fixed about the fifth century, remained almost rigid until the rediscovery of nature in the sixteenth. He was physician to Mithridates VI Eupator (120-63 B. C.), but his work was well known and appreciated at Rome, which became the place of resort for Greek talent.[84]

Celsus, who flourished about 20 B. C., wrote an excellent work on medicine, but gives all too little glimpse of anatomy and physiology. Rufus of Ephesus, however, in the next century practised dissection of apes and other animals. He described the decussation of the optic nerves and the capsule of the crystalline lens, and gave the first clear description that has survived of the structure of the eye. He regarded the nerves as originating from the brain, and distinguished between nerves of motion and of sensation. He described the oviduct of the sheep and rightly held that life was possible without the spleen.

The second Christian century brings us two writers who, while scientifically inconsiderable, acted as the main carriers of such tradition of Greek biology as reached the Middle Ages, Pliny and Dioscorides. Pliny (A. D. 23-79), though a Latin, owes almost everything of value in his encyclopaedia to Greek writings. In his *Natural History* we have a collection of current views on the nature, origin, and uses of plants and animals such as we might expect from an intelligent, industrious, and honest member of the landed class who was devoid of critical or special scientific skill. Scientifically the work is contemptible, but it demands mention in any study of the legacy of Greece, since it was, for centuries, a main conduit of the ancient teaching and observations on natural history. Read throughout the ages, alike in the darkest as in the more enlightened periods, copied and recopied, translated, commented on, extracted and abridged, a large part of Pliny's work has gradually passed into folk-keeping, so that through its agency the gipsy fortune-teller of to-day is still reciting garbled versions of the formulae of Aristotle and Hippocrates of two and a half millennia ago.

The fate of Dioscorides (flourished A. D. 60) has been not dissimilar. His work *On Materia Medica* consists of a series of short accounts of plants, arranged almost without reference to the nature of the plants themselves, but quite invaluable for its terse and striking descriptions which often include habits and habitats. Its history has shown it to be one of the most influential botanical treatises ever penned. It provided most of the little botanical knowledge that reached the Middle Ages. It furnished the chief stimulus to botanical research at the time of the Renaissance. It has decided the general form of every modern pharmacopœia. It has practically determined modern plant nomenclature both popular and scientific.

Translated into nearly every language from Anglo-Saxon and Provençal to Persian and Hebrew, appearing both abstracted and in full in innumerable beautifully illuminated manuscripts, some of which are still among the fairest treasures of the great national libraries, Dioscorides, the drug-monger, appealed to scholasticized minds for centuries. The frequency with which fragments of him are encountered in papyri shows how popular his work was in

Egypt in the third and fourth centuries. One of the earliest datable Greek codices in existence is a glorious volume of Dioscorides written in capitals,[85] thought worthy to form a wedding gift for a lady who was the daughter of one Roman emperor and the betrothed of a second.[86] The illustrations of this fifth century manuscript are a very valuable monument for the history of art and the chief adornment of what was once the Royal Library at Vienna[87] (figs. 9-10). Illustrated Latin translations of Dioscorides were in use in the time of Cassiodorus (490-585). A work based on it, similarly illustrated, but bearing the name of Apuleius, is among the most frequent of mediaeval botanical documents and the earliest surviving specimen is almost contemporary with Cassiodorus himself.[88] After the revival of learning Dioscorides continued to attract an immense amount of philological and botanical ability, and scores of editions of his works, many of them nobly illustrated, poured out of the presses of the sixteenth and seventeenth centuries.

Fifth century drawings from JULIANA ANICIA MS., copied from originals of first century B. C.(?)

Fig. 9.
ΣΟΝΚΟΣ ΤΡΥΦΕΡΟΣ = *Crepis paludosa, Mœn.*

Fig. 10.
ΓΕΡΑΝΙΟΝ = *Erodium malachoides, L.*

But the greatest biologist of the late Greek period, and indeed one of the greatest biologists of all time, was Claudius Galen of Pergamon (A. D. 131-201). Galen devoted himself to medicine from an early age, and in his twenty-first year we hear of him studying anatomy at Smyrna under Pelops. With the object of extending his knowledge of drugs he early made long journeys to Asia Minor. Later he proceeded to Alexandria, where he improved his anatomical equipment, and here, he tells us, he examined a human skeleton. It is indeed probable that his direct practical acquaintance with human anatomy was limited to the skeleton and that dissection of the human body was no longer carried on at Alexandria in his time. Thus his physiology and anatomy had to be derived mainly from animal sources. He is the most voluminous of all ancient scientific writers and one of the most voluminous writers of antiquity in any department. We are not here concerned with the medical material which mainly fills these huge volumes, but merely with the physiological views which not only prevailed in medicine until Harvey and after, but also governed for fifteen hundred years alike the scientific and the popular ideas on the nature and workings of the animal body, and have for centuries been embedded in our speech. A knowledge of these physiological views of Galen is necessary for any understanding of the history of biology and illuminates many literary allusions of the Middle Ages and Renaissance.

Between the foundation of the Alexandrian school and the time of Galen, medicine was divided among a great number of sects. Galen was an eclectic and took portions of his teaching from many of these schools, but he was also a naturalist of great ability and industry, and knew well the value of the experimental way. Yet he was a somewhat windy philosopher and, priding himself on his philosophic powers, did not hesitate to draw conclusions from evidence which was by no means always adequate. The physiological system that he thus succeeded in building up we may now briefly consider (<u>fig. 11</u>).

The basic principle of life, in the Galenic physiology, is a *spirit, anima* or *pneuma*, drawn from the general world-soul in the act of respiration. It enters the body through the *rough artery* (τραχεῖα ἀρτηρία, *arteria aspera* of mediaeval notation), the organ known to our nomenclature as the trachea. From this trachea the pneuma passes to the lung and then, through the *vein-like artery* (ἀρτηρία φλεβώδης, *arteria venalis* of mediaeval writers, the pulmonary vein of our nomenclature), to the left ventricle. Here it will be best to leave it for a moment and trace the vascular system along a different route.

Ingested food, passing down the alimentary tract, was absorbed as chyle from the intestine, collected by the portal vessel, and conveyed by it to the liver. That organ, the site of the innate heat in Galen's view, had the power of elaborating the chyle into venous blood and of imbuing it with a spirit or pneuma which is innate in all living substance, so long as it remains alive, the *natural spirits* (πνεῦμα φυσικόν, *spiritus naturalis* of the mediaevals). Charged with this, and also with the nutritive material derived from the food, the venous blood is distributed by the liver through the veins which arise from it in the same way as the arteries from the heart. These veins carry nourishment and *natural spirits* to all parts of the body. *Iecur fons venarum*, the liver as the source of the veins, remained through the centuries the watchword of the Galenic physiology. The blood was held to ebb and flow continuously in the veins during life.

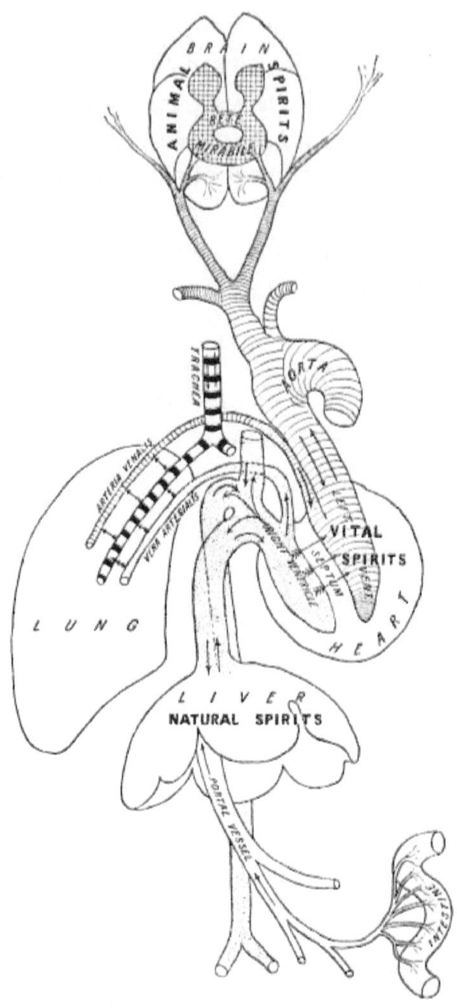

FIG. 11. Illustrating Galen's physiological teaching.

Now from the liver arose one great vessel, the hepatic vein, from division of which the others were held to come off as branches. Of these branches, one, our *common vena cava*, entered the right side of the heart. For the blood that it conveyed to the heart there were two fates possible. The greater part remained awhile in the ventricle, parting with its impurities and vapours, exhalations of the organs, which were carried off by the *artery-like vein* (φλὲπς ἀρτηριώδης, the mediaeval *vena pulmonalis*, our pulmonary artery) to the lung and then exhaled to the outer air. These impurities and vapours gave its poisonous and suffocating character to the breath. Having parted thus with its impurities, the venous blood ebbed back again from the right ventricle into the venous system. But for a small fraction of the venous blood that entered the right ventricle another fate was reserved. This small fraction of venous blood, charged still with the *natural spirits* derived from the liver, passed through minute channels in the septum between the ventricles and entered the left chamber. Arrived there, it encountered the external pneuma and became thereby elaborated into a higher form of spirit, the *vital spirits* (πνεῦμα ζωτικόν, *spiritus vitalis*), which is distributed together with blood by the arterial system to various parts of the body. In the arterial system it also ebbed and flowed, and might be seen and felt to pulsate there.

But among the great arterial vessels that sent forth arterial blood thus charged with vital spirits were certain vessels which ascended to the brain. Before reaching that organ they divided up into minute channels, the *rete mirabile* (πλέγμα μέγιστον θαῦμα), and passing into the brain became converted by the action of that organ into a yet higher type of spirits, the *animal spirits* (πνεῦμα ψυχικόν, *spiritus animalis*), an ethereal substance distributed to the various parts of the body by the structures known to-day as nerves, but believed then to be hollow channels. The three fundamental faculties (δυνάμεις), the *natural*, the *vital*, and the *animal*, which brought into action the corresponding functions of the body, thus originated as an expression of the primal force or pneuma.

This physiology, we may emphasize, is not derived from an investigation of human anatomy. In the human brain there is no *rete mirabile*, though such an organ is found in the calf.

In the human liver there is no *hepatic vein*, though such an organ is found in the dog. Dogs, calves, pigs, bears, and, above all, Barbary apes were freely dissected by Galen and were the creatures from which he derived his physiological ideas. Many of Galen's anatomical and physiological errors are due to his attributing to one creature the structures found in another, a fact that only very gradually dawned on the Renaissance anatomists.

The whole knowledge possessed by the world in the department of physiology from the third to the seventeenth century, nearly all the biological conceptions till the thirteenth, and most of the anatomy and much of the botany until the sixteenth century, all the ideas of the physical structure of living things throughout the Middle Ages, were contained in a small number of these works of Galen. The biological works of Aristotle and Theophrastus lingered precariously in a few rare manuscripts in the monasteries of the East; the total output of hundreds of years of Alexandrian and Pergamenian activities was utterly destroyed; the Ionian biological works, of which a sample has by a miracle survived, were forgotten; but these vast, windy, ill-arranged treatises of Galen lingered on. Translated into Latin, Syriac, Arabic, and Hebrew, they saturated the intellectual world of the Middle Ages. Commented on by later Greek writers, who were themselves in turn translated into the same list of languages, they were yet again served up under the names of such Greek writers as Oribasius, Paul of Aegina, or Alexander of Tralles.

What is the secret of the vitality of these Galenic biological conceptions? The answer can be given in four words. *Galen is a teleologist*; and a teleologist of a kind whose views happened to fit in with the prevailing theological attitude of the Middle Ages, whether Christian, Moslem, or Jewish. According to him everything which exists and displays activity in the human body originates in and is formed by an intelligent being and on an intelligent plan, so that the organ in structure and function is the result of that plan. 'It was the Creator's infinite wisdom which selected the best means to attain his beneficent ends, and it is a proof of His omnipotence that he created every good thing according to His design, and thereby fulfilled His will.'[89]

After Galen there is a thousand years of darkness, and biology ceases to have a history. The mind of the Dark Ages turned towards theology, and such remains of Neoplatonic philosophy as were absorbed into the religious system were little likely to be of aid to the scientific attitude. One department of positive knowledge must of course persist. Men still suffered from the infirmities of the flesh and still sought relief from them. But the books from which that advice was sought had nothing to do with general principles nor with knowledge as such. They were the most wretched of the treatises that still masqueraded under the names of Hippocrates and Galen, mostly mere formularies, antidotaries, or perhaps at best symptom lists. And, when the depression of the western intellect had passed its worst, there was still no biological material on which it could be nourished.

The prevailing interest of the barbarian world, at last beginning to settle into its heritage of antiquity, was with Logic. Of Aristotle there survived in Latin dress only the *Categories* and the *De interpretatione*, the merciful legacy of Boethius, the last of the philosophers. Had a translation of Aristotle's *Historia animalium* or *De generatione animalium* survived, had a Latin version of the Hippocratic work *On generation* or of the treatises of Theophrastus *On plants* reached the earlier Middle Ages, the whole mental history of Europe might have been different and the rediscovery of nature might have been antedated by centuries. But this was a change of heart for which the world had long to wait; something much less was the earliest biological gift of Greece. The gift, when it came, came in two forms, one of which has not been adequately recognized, but both are equally her legacy. These two forms are, firstly, the well-known work of the early translators and, secondly, the tardily recognized work of certain schools of minor art.

The earliest biological treatises that became accessible in the west were rendered not from Greek but from Arabic. The first of them was perhaps the treatise περὶ μυῶν κινήσεως, *On movement of muscles* of Galen, a work which contains more than its title suggests and indeed sets forth much of the Galenic physiological system. It was rendered into Latin from the Arabic of Joannitius (Hunain ibn Ishaq, 809-73), probably about the year 1200, by one Mark of Toledo. It attracted little attention, but very soon after biological works of Aristotle began to become accessible. The first was probably the fragment *On plants*. The Greek original of this is lost, and besides the Latin, only an Arabic version of a former Arabic translation of a Syriac rendering of a Greek commentary is now known! Such a work appeared from the hand of a translator known as Alfred the Englishman about 1220 or a little later. Neither it nor another work from the same translator, *On the motion of the heart*, which sought to establish the primacy of that organ on Aristotelian grounds, can be said to contain any of the spirit of the master.[90]

A little better than these is the work of the wizard Michael the Scot (1175?-1234?). Roger Bacon tells us that Michael in 1230 'appeared [at Oxford], bringing with him the works of Aristotle in natural history and mathematics, with wise expositors, so that the philosophy of Aristotle was magnified among the Latins'. Scott produced his work *De animalibus* about this date and he included in it the three great biological works of Aristotle, all rendered from an inferior Arabic version.[91] Albertus Magnus (1206-80) had not as yet a translation direct from the Greek to go upon for his great commentary on the *History of animals*, but he depended on Scott. The biological works of Aristotle were rendered into Latin direct from the Greek in the year 1260 probably by William of Moerbeke.[92] Such translations, appearing in the full scholastic age when everything was against direct observation, cannot be said to have fallen on a fertile ground. They presented an ordered account of nature and a good method of investigation, but these were gifts to a society that knew little of their real value.[93]

Yet the advent of these texts was coincident with a returning desire to observe nature. Albert, with all his scholasticism, was no contemptible naturalist. He may be said to have begun first-hand plant study in modern times so far as literary records are concerned. His book *De vegetabilibus* contains excellent observations, and he is worthy of inclusion among the fathers of botany. In his vast treatise *De animalibus*, hampered as he is by his learning and verbosity, he shows himself a true observer and one who has absorbed something of the spirit of the great naturalist to whose works he had devoted a lifetime of study and on which he professes to be commenting. We see clearly the leaven of the Aristotelian spirit working, though Albert is still a schoolman. We may select for quotation a passage on the generation of fish, a subject on which some of Aristotle's most remarkable descriptions remained unconfirmed till modern times. These descriptions impressed Albert in the same way as they do the modern naturalist. To those who know nothing of the stimulating power of the Aristotelian biological works, Albert's description of the embryos of fish and his accurate distinction of their mode of development from that of birds, by the absence of an allantoic membrane in the one and its presence in the other, must surely be startling. Albert depends on Aristotle—a third-hand version of Aristotle—but does not slavishly follow him.

'Between the mode of development (*anathomiam generationis*) of birds' and fishes' eggs there is this difference: during the development of the fish the second of the two veins which extend from the heart [as described by Aristotle in birds] does not exist. For we do not find the vein which extends to the outer covering in the eggs of birds which some wrongly call the navel because it carries the blood to the exterior parts; but we do find the vein that corresponds to the yolk vein of birds, for this vein imbibes the nourishment by which the limbs increase.... In fishes as in birds, channels extend from the heart first to the head and the eyes, and first in them appear the great upper parts. As the growth of the young fish increases the albumen decreases, being incorporated into the members of the young fish, and it disappears entirely when development and formation are complete. The beating of the heart ... is conveyed to the lower part of the belly, carrying pulse and life to the inferior members.

'While the young [fish] are small and not yet fully developed they have veins of great length which take the place of the navel-string, but as they grow and develop, these shorten and contract into the body towards the heart, as we have said about birds. The young fish and the eggs are enclosed and in a covering, as are the eggs and young of birds. This covering resembles the dura mater [of the brain], and beneath it is another [corresponding therefore to the pia mater of the brain] which contains the young animal and nothing else.'[94]

In the next century Conrad von Megenberg (1309-98) produced his *Book of Nature*, a complete work on natural history, the first of the kind in the vernacular, founded on Latin versions, now rendered direct from the Greek, of the Aristotelian and Galenic biological works. It is well ordered and opens with a systematic account of the structure and physiology of man as a type of the animal creation, which is then systematically described and followed by an account of plants. Conrad, though guided by Aristotle, uses his own eyes and ears, and with him and Albert the era of direct observation has begun.[95]

But there was another department in which the legacy of Greece found an even earlier appreciation. For centuries the illustrations to herbals and bestiaries had been copied from hand to hand, continuing a tradition that had its rise with Greek artists of the first century B. C. But their work, copied at each stage without reference to the object, moved constantly farther from resemblance to the original. At last the illustrations became little but formal patterns, a state in which they remained in some late copies prepared as recently as the sixteenth century. But at a certain period a change set in, and the artist, no longer content to rely on tradition, appeals at last to nature. This new stirring in art corresponds with the new stirring in letters, the Arabian revival—itself a legacy of Greece, though sadly deteriorated in transit—that gave rise to scholasticism. In much of the beautiful carved and sculptured work of the French cathedrals the new movement appears in the earlier part of the thirteenth century. At such a place as Chartres we see the attempt to render plants and animals faithfully in stone as early as 1240 or before. In the easier medium of parchment the same tendency appears even earlier. When once it begins the process progresses slowly until the great recovery of the Greek texts in the fifteenth century, when it is again accelerated.

During the sixteenth century the energy of botanists and zoologists was largely absorbed in producing most carefully annotated and illustrated editions of Dioscorides and Theophrastus and accounts of animals, habits, and structure that were intended to illustrate the writings of Aristotle, while the anatomists explored the bodies of man and beast to confirm or refute Galen. The great monographs on birds, fishes, and plants of this period, ostensibly little but commentaries on Pliny, Aristotle, and Dioscorides, represent really the first important efforts of modern times at a natural history. They pass naturally into the encyclopaedias of the later sixteenth century, and these into the physiological works of the seventeenth. Aristotle was never a dead hand in Biology as he was in Physics, and this for the reason that he was a great biologist but was not a great physicist.

With the advance of the sixteenth century the works of Aristotle, and to a less extent those of Dioscorides and Galen, became the great stimulus to the foundation of a new biological science. Matthioli (1520-77), in his commentary on Dioscorides (first edition 1544), which was one of the first works of its type to appear in the vernacular, made a number of first-hand observations on the habits and structure of plants that is startling even to a modern botanist. About the same time Galenic physiology, expressed also in numerous works in the vulgar tongue and rousing the curiosity of the physicians, became the clear parent of modern physiology and comparative anatomy. But, above all, the Aristotelian biological works were fertilizers of the mind. It is very interesting to watch a fine observer such as Fabricius ab Acquapendente (1537-1619) laying the foundations of modern embryology in a splendid series of first-hand observations, treating his own great researches almost as a commentary on Aristotle. What an impressive contrast to the arid physics of the time based also on Aristotle! 'My purpose', says Fabricius, 'is to treat of the formation of the foetus in every animal, setting out from that which proceeds from the egg: for this ought to take precedence of all other discussion of the subject, both because it is not difficult to make out Aristotle's view of the matter, and because his treatise on the Formation of the Foetus from the egg is by far the fullest, and the subject is by much the most extensive and difficult.'[96]

The industrious and careful Fabricius, with a wonderful talent for observation lit not by his own lamp but by that of Aristotle, bears a relation to the master much like that held by Aristotle's pupil in the flesh, Theophrastus. The works of the two men, Fabricius and Theophrastus, bear indeed a resemblance to each other. Both rely on the same group of general ideas, both progress in much the same ordered calm from observation to observation, both have an inspiration which is efficient and stimulating but below the greatest, both are enthusiastic and effective as investigators of fact, but timid and ineffective in drawing conclusions.

But Fabricius was more happy in his pupils than Theophrastus, for we may watch the same Aristotelian ideas fermenting in the mind of Fabricius's successor, the greatest biologist since Aristotle himself, William Harvey (1578-1657).[97] This writer's work *On generation* is a careful commentary on Aristotle's work on the same topic, but it is a commentary not in the old sense but in the spirit of Aristotle himself. Each statement is weighed and tested in the light of experience, and the younger naturalist, with all his reverence for Aristotle, does not hesitate to criticize his conclusions. He exhibits an independence of thought, an ingenuity in experiment, and a power of deduction that places his treatise as the middle term of the three great works on embryology of which the other members are those of Aristotle and Karl Ernst von Baer (1796-1870).[98]

With the second half of the seventeenth century and during a large part of the eighteenth the biological works of Aristotle attracted less attention. The battle against the Aristotelian physics had been fought and won, but with them the biological works of Aristotle unjustly passed into the shadow that overhung all the idols of the Middle Ages.

The rediscovery of the Aristotelian biology is a modern thing. The collection of the vast wealth of living forms absorbed the energies of the generations of naturalists from

Ray (1627-1705) and Willoughby (1635-72) to Réaumur (1683-1757) and Linnaeus (1707-1778) and beyond to the nineteenth century. The magnitude and fascination of the work seems almost to have excluded general ideas. With the end of this period and the advent of a more philosophical type of naturalist, such as Cuvier (1769-1832) and members of the Saint-Hilaire family, Aristotle came again to his own. Since the dawn of the nineteenth century, and since naturalists have been in a position to verify the work of Aristotle, his reputation as a naturalist has continuously risen. Johannes Müller (1801-58), Richard Owen (1804-92), George Henry Lewes (1817-78), William Ogle (1827-1912) are a few of the long line of those who have derived direct inspiration from his biological work. With improved modern methods of investigation the problems of generation have absorbed a large amount of biological attention, and interest has become specially concentrated on Aristotle's work on that topic which is perhaps, at the moment, more widely read than any biological treatise, ancient or modern, except the works of Darwin. That great naturalist wrote to Ogle in 1882: 'From quotations I had seen I had a high notion of Aristotle's merits, but I had not the most remote notion what a wonderful man he was. Linnaeus and Cuvier have been my two gods, though in very different ways, but they were mere schoolboys to old Aristotle.'

GREEK MEDICINE

Ἡρόφιλος δὲ Ἦεν τῷ
Διαιτητικῷ καὶ σοφίαν φησὶν
ἀνεπίδεικτον καὶ τέχνην ἄδηλον
καὶ ἰσχὺν ἀναγώνιστον καὶ
πλοῦτον ἀχρεῖον καὶ λόγον
ἀδύνατον, ὑγιείας ἀπούσης.

Herophilus, a Greek philosopher and physician (*c.* 300 B. C.), has truly written 'that Science and Art have equally nothing to show, that Strength is incapable of effort, Wealth useless, and Eloquence powerless if Health be wanting'.[99] All peoples therefore have had their methods of treating those departures from health that we call disease, and among peoples of higher culture such methods have been reduced in most cases to something resembling a system. In antiquity, as now, a variety of such systems were in vogue, and those nations who practised the art of writing from an early date have left considerable records of their medical methods and doctrines. We may thus form a fairly good idea of the medical principles of the Mesopotamian, the Egyptian, the Iranian, the Indian, and the Chinese civilizations. Much in these systems, as in the medical procedure of more primitive tribes, was based upon some theory of disease which fitted in with a larger theory of the nature of evil. Of these theories the commonest was and is the demonic, the view that regards deviation from the normal state of health as due either to the attacks of supernatural beings or to their actual entry into the body of the sufferer. A medical system based on such a view is susceptible of great elaboration in a higher civilization, but not being founded on observation is hardly capable of indefinite development, for a point must ultimately be reached at which the mind recoils from complex conclusions far remote from observed phenomena. The medicine of the ancient and settled civilization of such a people as the Assyro-Babylonians, for instance, of which substantial traces have been recovered, is hardly, if at all, more effective, though far more systematized, than that of many a wild and unlettered tribe that may be observed to-day. Of such medicine as this we may give an account, but we can hardly write a *history*. We cannot establish those elements of continuity and of development from which alone history can be constructed.

It is the distinction of the Greeks alone among the nations of antiquity that they practised a system of medicine based not on theory but on observation accumulated systematically as time went on. The claim can be made for the Greeks that some at least among them were deflected by no theory, were deceived by no theurgy, were hampered by no tradition in their search for the facts of disease and in their attempts at interpreting its phenomena. Only the Greeks among the ancients could look on their healers as *physicians* (= naturalists, φύσις = nature), and that word itself stands as a lasting reminder of their achievement.[100]

At a certain stage in the history of the Western world—the exact point in time may be disputed but the event is admitted by all—men turned to explore the treasures of the ancient wisdom and the whole mass of Greek medical learning was gradually laid before the student. That mass contained much dross, material that survived from early as from late Greek times which was hardly, if at all, superior to the debased compositions that circulated in the name of medicine in the middle centuries. But the recovered Greek medical writings also contained some material of the purest and most scientific type, and that material and the spirit in which it was written, form the debt of modern medicine to antiquity.

It is a debt the value of which cannot be exaggerated. The physicians of the revival of learning, and for long after, doubtless pinned their faith too much to the written word of their Greek forbears and sought to imprison the free spirit of Hippocrates and Galen in the rigid wall of their own rediscovered texts. The great medical pioneers of a somewhat later age, enraged by this attempt, the real nature of which was largely hidden from them, not infrequently revolted and rightly revolted against the bondage to the Greeks in which they had been brought up. Yet it is sure that these modern discoverers were the true inheritors of the Greeks. Without Herophilus we should have had no Harvey and the rise of physiology might have been delayed for centuries; had Galen's works not survived, Vesalius would never have reconstructed Anatomy, and Surgery too might have stayed behind with her laggard sister, Medicine; the Hippocratic collection was the necessary and acknowledged basis for the work of the greatest of modern clinical observers, Thomas Sydenham, and the teaching of Hippocrates and of his school is the substantial basis of instruction in the wards of a modern hospital. In the pages which follow we propose therefore to review the general character of medical knowledge in the best Greek period and to consider briefly how much of that great heritage remained accessible to the earlier modern physicians. The reader will thus be able to form some estimate of the degree to which the legacy has been passed on to our own times.

It is evident that among such a group of peoples as the Greeks, varying in state of civilization, in mental power, in geographical and economic position and in general outlook, the practice of medicine can have been by no means uniform. Without any method of centralizing medical education and standardizing teaching there was a great variety of doctrines and of practice in vogue among them, and much of this was on a low level of folk custom. Such lower grade material of Greek origin has come down to us in abundance, though much of it, curiously enough, from a later time. But the overwhelming mass of earlier Greek medical literature sets forth for us a pure scientific effort to observe and to classify disease, to make generalizations from carefully collected data, to explain the origin of disease on rational grounds, and to apply remedies, when possible, on a reasoned basis. We may thus rest fairly well assured that, despite serious and irreparable losses, we are still in possession of some of the very finest products of the Greek medical intellect.

There is ample evidence that the Greeks inherited, in common with many other peoples of Mediterranean and Asiatic origin, a whole system of magical or at least non-rational pharmacy and medicine from a remoter ancestry. Striking parallels can be drawn between these folk elements among the Greeks and the medical systems of the early Romans, as well as with the medicine of the Indian Vedas, of the ancient Egyptians, and of the earliest European barbarian writings. It is thus reasonable to suppose that these elements, when they appear in later Greek writings, represent more primitive folk elements working up, under the influence of social disintegration and consequent mental deterioration, through the upper strata of the literate Greek world. But with these elements, intensely interesting to the anthropologist, the psychologist, the ethnologist, and to the historian of religion, we are not here greatly concerned. Important as they are, they constitute no part of the special claim of the Greek people to distinction, but rather aid us in uniting the Greek mentality with that of other kindred peoples. Here we shall rather discuss the course of Greek scientific medicine proper, the type of medical doctrine and practice, capable of development in the proper sense of the word, that forms the basis of our modern system. We are concerned, in fact, with the earliest evolutionary medicine.

We need hardly discuss the first origins of Greek Medicine. The material is scanty and the conclusions somewhat doubtful and perhaps premature, for the discovery of a considerable fragment of the historical work of Menon, a pupil of Aristotle, containing a description of the views of some of the precursors of the Hippocratic school, renews a hope that more extended investigation may yield further information as to the sources and nature of the earliest Greek medical writings.[101] The study of Mesopotamian star-lore has linked it up with early Greek astronomical science. The efforts of cuneiform scholars have not, however, been equally successful for medicine, and on the whole the general tendency of modern research is to give less weight to Mesopotamian and more to Egyptian sources than had previously been admitted; thus, as an instance, some prescriptions in the Ebers papyrus of the eighteenth dynasty (about the sixteenth century B. C.) discovered at Thebes in 1872 resemble certain formulae in the Corpus Hippocraticum. A number of drugs, too, habitually used by the Greeks, such as *Andropogon, Cardamoms*, and *Sesame orientalis*, are of Indian origin. There are also the Minoan cultures to be considered, and our knowledge is not yet sufficient to speak of the heritage that Greek medicine may or may not have derived from that source, though it seems not improbable that Greek hygiene may here owe a debt.[102] Omitting, therefore, this early epoch, we pass direct to the later period, between the sixth and fourth centuries, from which documents have actually come down to us.

The earliest medical school of which we have definite information is that of Cnidus, a Lacedaemonian colony in Asiatic Doris. Its origin may perhaps reach back to the seventh century B. C. We have actual records that the teachers of Cnidus were accustomed to collect systematically the phenomena of disease, of which they had produced a very complex classification, and we probably possess also several of their actual works. The physicians of Cos, their only contemporary critics whose writings have survived, considered that the Cnidian physicians paid too much attention to the actual sensations of the patient and to the physical signs of the disease. The most important of the Cnidian doctrines were drawn up in a series of *Sentences* or Aphorisms, and these, it appears, inculcated a treatment along Egyptian lines of the symptom or at most the disease, rather than the patient, a statement borne out by the contents of the gynaecological works of probable Cnidian origin included in the so-called 'Hippocratic Collection'. A few names of Cnidian physicians have, moreover, come down to us with titles of their works, and a later statement that they practised anatomy. There can be little doubt too that the Cnidian school drew also on Persian and Indian Medicine.

The origin of the school of the neighbouring island of Cos was a little later than that of Cnidus and probably dates from the sixth century B. C. Of the Coan school, or at least of the general tendencies that it represented, we have a magnificent and copious literary monument in the Corpus Hippocraticum, a collection which was probably put together in the early part of the third century B. C. by a commission of Alexandrian scholars at the order of the book-loving Ptolemy Soter (reigned 323-285 B. C.). The elements of which this collection is composed are of varying dates from the sixth to the fourth century B. C., and of varying value and origin, but they mainly represent the point of view of physicians of the eastern part of the Greek world in the fifth and fourth centuries.

The most obvious feature, the outstanding element that at once strikes the modern observer in these 'Coan' writings, is the enormous emphasis laid on the actual course of disease. 'It appears to me a most excellent thing', so opens one of the greatest of the Hippocratic works, 'for a physician to cultivate *pronoia*.[103] Foreknowing and foretelling in the presence of the sick the past, present, and future (of their symptoms) and explaining all that the patients are neglecting, he would be believed to understand their condition, so that men would have confidence to entrust themselves to his care.... Thus he would win just respect and be a good physician. By an earlier forecast in each case he would be more able to tend those aright who have a chance of surviving, and by foreseeing and stating who will die, and who will survive, he will escape blame....'[104]

Just as the Cnidians by dividing up diseases according to symptoms over-emphasized diagnosis and over-elaborated treatment, so the Coans laid very great force on prognosis and adopted therefore a largely expectant attitude towards diseases. Both Cnidian and Coan physicians were held together by a common bond which was, historically if not actually, related to temple worship. Physicians leagued together in the name of a god, as were the Asclepiadae, might escape, and did escape, the baser theurgic elements of temple medicine. Of these they were as devoid as a modern Catholic physician might be expected to be free from the absurdities of Lourdes. But the extreme cult of prognosis among the Coans may not improbably be traced back to the medical lore of the temple soothsayers whose divine omens were replaced by indications of a physical nature in the patient himself.[105] We are tempted too to link it with that process of astronomical and astrological prognosis practised in the Mesopotamian civilizations from which Ionia imitated and derived so much. Religion had thus the same relation to medicine that it would have with a modern 'religious' medical man as suggesting the motive and determining the general direction of his practice though without influence on the details and method.

During the development of the Coan medical school along these lines in the sixth and fifth centuries, there was going on a most remarkable movement at the very other extreme of the Greek world. Into the course and general importance of Sicilian philosophy it is not our place to enter, but that extraordinary movement was not without its repercussion on medical theory and practice. Very important in this direction was Empedocles of Agrigentum (*c.* 500-*c.* 430 B. C.). His view that the blood is the seat of the 'innate heat', ἔμφυτον θερμόν, he took from folk belief—'the blood is the life'—and this innate heat he closely identified with soul. More profitable was his doctrine that breathing takes place not only through what are now known as the respiratory passages but also through the pores of the skin. His teaching led to a belief in the heart as the centre of the vascular system and the chief organ of the 'pneuma' which was distributed by the blood vessels. This pneuma was equivalent to both soul and life, but it was something more. It was identified with air and breath, and the pneuma could be seen to rise as shimmering steam from the shed blood of the sacrificial victim—for was not the blood its natural home? There was a pneuma, too, that interpenetrated the universe around us and gave it those qualities of life that it was felt to possess. Anaximenes (*c.* 610-*c.* 545 B. C.), an Ionian predecessor of Empedocles, may be said to have defined for us these functions of the pneuma; οἷον ἡ ψυχὴ ἡ ἡμετέρα ἀὴρ οὖσα συγκρατεί ἡμᾶς, ὅλον τὸν κόσμον πνεῦμα καὶ ἀὴρ περιέχει 'As our soul, being air, sustains us, so pneuma and air pervade the whole universe';[106] but it is the speculation of Empedocles himself that came to be regarded as the basis of the Pneumatic School in Medicine which had later very important developments.

Another early member of the Western school who made important contributions to medical doctrine—in which relation alone we need consider him—was Pythagoras of Samos (*c.* 580-*c.* 490 B. C.). For him number, as the purest conception, formed the basis of philosophy. Unity was the symbol of perfection and corresponded to God Himself. The material universe was represented by 2, and was divided by the number 12, whence we have 3 worlds and 4 spheres. These in turn, according at least to the later Pythagoreans, give rise to the four elements, earth, air, fire, and water—a primary doctrine of medicine and of science derived perhaps from ancient Egypt and surviving for more than two millennia. The Pythagoreans taught, too, of the existence of an animal soul, an emanation of the soul of the universe. In all this we may distinguish the germ of that doctrine of the relation of man and universe, microcosm and macrocosm, which, suppressed as irrelevant in the Hippocratic works, reappears in the Platonic and especially in the Neoplatonic writings, and forms a very important dogma in later medicine.

A pupil of Pythagoras and an older contemporary of Empedocles was Alcmaeon of Croton (*c.* 500 B. C.), who began to construct a positive basis for medical science by the practice of dissection of animals, and discovered the optic nerves and the Eustachian tubes. He even extended his researches to Embryology, describing the head of the foetus as the first part to be developed—a justifiable deduction from appearances. Alcmaeon introduced also the doctrine that health depends on harmony, disease on discord of the elements within the body. Curiosity as to the distribution of the vessels was excited by Empedocles and Alcmaeon and led to further dissection, and Alcmaeon's pupils Acron (*c.* 480 B. C.) and Pausanias (*c.* 480 B. C.), and the later Philistion of Lokri,[107] the contemporary of Plato, all made anatomical investigations.

The views of Empedocles, and especially his doctrine that regarded the heart as the main site of the pneuma, though rejected by the Coan school as a whole, were not without influence on Ionia. Diogenes of Apollonia, the philosopher of pneumatism, a late fifth century writer who must have been contemporary with Hippocrates the Great, himself made an investigation of the blood vessels; and the influence of the same school may be traced in a little work περὶ καρδίης, *On the heart*, which is the best anatomical treatise of the Hippocratic Collection. This work describes the aorta and the pulmonary artery as well as the three valves at the root of each of the great vessels, and it speaks of experiments to test their validity. It treats of the pericardium and of the pericardial fluid and perhaps of the musculi papillares, and contrasts the thickness of the walls of right and left   ventricles. The author considers that the left ventricle is empty of blood—as indeed it is after death—and is the source of the innate heat and of the absolute intelligence. These views fit in with the doctrines of Empedocles, so that we may perhaps even venture to regard this work as a surviving document of the Sicilian school. It is interesting to observe that we have here the first hint of human dissection, for the author tells us that the hearts of animals may be compared to that of man. The distinction of having been the first to write on human anatomy, as such, belongs however, probably to a later writer, Diocles, son of Archidamus of Carystus, who lived in the fourth century B. C.[108]

We may now turn to the Hippocratic Corpus as a whole. This collection consists of about 60 or 70 separate works, written at various periods and in various states of preservation. At best only a very small proportion of them can be attributed to Hippocrates, but the discussion of the general question of the 'genuineness' of the works is now admitted to be futile, for it is certain that we have no criteria whatever to determine whether or no a particular work be from the pen of the Father of Medicine, and the most we can ever say of such a treatise is that it appears to be of his school and in his spirit. Yet among the great gifts of this collection to our time and to all time are two which stand out above all others, the picture of a man, and the picture of a method.

The man is Hippocrates himself. Of the actual details of his life we know next to nothing. His period of greatest activity falls about 400 B. C. He seems to have led a wandering life. Born of a long line of physicians in the island of Cos, he exerted his activities in Thrace, Abdera, Delos, the Propontis (Cyzicus), Thasos, Thessaly (notably at Larissa and Meliboea), Athens, and elsewhere, dying at Larissa in extreme old age about the year 377 B. C. He had many pupils, among whom were his two sons Thessalus and Dracon, who also undertook journeys, his son-in-law Polybus, of whose works a fragment has been preserved for us by Aristotle,[109] together with three other Coans bearing the names Apollonius, Dexippus, and Praxagoras. This is practically all we know of him with certainty. But though this glimpse is very dim and distant, yet we cannot exaggerate the influence on the course of medicine and the value for physicians of all time of the traditional picture that was early formed of him and that may indeed well be drawn again from the works bearing his name. In beauty and dignity that figure is beyond praise. Perhaps gaining in stateliness what he loses in clearness, Hippocrates will ever remain the type of the perfect physician. Learned, observant, humane, with a profound reverence for the claims of his patients, but an overmastering desire that his experience shall benefit others, orderly and calm, disturbed only by anxiety to record his knowledge for the use of his brother physicians and for the relief of suffering, grave, thoughtful and reticent, pure of mind and master of his passions, this is no overdrawn picture of the Father of Medicine as he appeared to his contemporaries and successors. It is a figure of character and virtue which has had an ethical value to medical men of all ages comparable only to the influence exerted on their followers by the founders of the great religions. If one needed a maxim to place upon the statue of Hippocrates, none could be found better than that from the book Παραγγελίαι, *Precepts*:

ἢν γὰρ παρῇ φιλανθρωπίη πάρεστι καὶ φιλοτεχνίη 'Where the love of man is, there also is love of the Art.'[110]

Fig. 1.   HIPPOCRATES

British Museum, second or third century B. C.

Fig. 2.   ASCLEPIUS

British Museum, fourth century B. C.

The numerous busts of him which have reached our time are no portraits. But the best of them are something much better and more helpful to us than any portrait. They are idealized representations of the kind of man a physician should be and was in the eyes of the best and wisest of the Greeks. (<u>See Fig. 1.</u>)

The method of the Hippocratic writers is that known to-day as the 'inductive'. Without the vast scientific heritage that is in our own hands, with only a comparatively small number of observations drawn from the Coan and neighbouring schools, surrounded by all manner of bizarre oriental religions in which no adequate relation of cause and effect was recognized, and above all constantly urged by the exuberant genius for speculation of that Greek people in the midst of whom they lived and whose intellectual temptations they shared, they remain nevertheless, for the most part, patient observers of fact, sceptical of the marvellous and the unverifiable, hesitating to theorize beyond the data, yet eager always to generalize from actual experience; calm, faithful, effective servants of the sick. There is almost no type of mental activity known to us that was not exhibited by the Greeks and cannot be paralleled from their writings; but careful and constant return to verification from experience, expressed in a record of actual observations— the habitual method adopted in modern scientific departments—is rare among them except in these early medical authors.

The spirit of their practice cannot be better illustrated than by the words of the so-called 'Hippocratic oath':

'I swear by Apollo the healer,
and Asclepius, and Hygieia, and
All-heal (Panacea) and all the gods
and goddesses ... that, according to
my ability and judgement, I will
keep this Oath and this
stipulation—to reckon him who
taught me this Art as dear to me as
those who bore me ... to look upon
his offspring as my own brothers,
and to teach them this Art, if they
would learn it, without fee or
stipulation. By precept, lecture,
and all other modes of instruction,
I will impart a knowledge of the
Art to my own sons, and those of
my teacher, and to disciples bound
by a stipulation and oath according
to the Law of Medicine, but to

none other. I will follow that system of regimen which, according to my ability and judgement, I consider for the benefit of my patients, and abstain from whatever is deleterious and mischievous. I will give no deadly medicine to any one if asked, nor suggest any such counsel; nor will I aid a woman to produce abortion. With purity and holiness I will pass my life and practise my Art.... Into whatever houses I enter, I will go there for the benefit of the sick, and will abstain from every act of mischief and corruption; and above all from seduction.... Whatever in my professional practice—or even not in connexion with it—I see or hear in the lives of men which ought not to be spoken of abroad, I will not divulge, deeming that on such matters we should be silent. While I keep this Oath unviolated, may it be granted me to enjoy life and the practice of the Art, always respected among men, but should I break or violate this Oath, may the reverse be my lot.'

Respected equally throughout the ages by Arab, Jew, and Christian, the oath remains the watchword of the profession of medicine.[111] The ethical value of such a declaration could not escape the attention even of a Byzantine formalist, and it is interesting to observe that in our oldest Greek manuscript of the Hippocratic text, dating from the tenth century, this magnificent passage is headed by the words 'from the oath of Hippocrates according as it may be sworn by a Christian.'[112]

When we examine the Hippocratic corpus more closely, we discern that not only are the treatises by many hands, but there is not even a uniform opinion and doctrine running through them. This is well brought out by some of the more famous of the phrases of this remarkable collection. Thus a well-known passage from the *Airs, Waters, and Places* tells us that the Scythians attribute a certain physical disability to a god, 'but it appears to me', says the author, 'that these affections are just as much divine as are all others and that no disease is either more divine or more human than another, but that all are equally divine, for each of them has its own nature, and none of them arise without a natural cause.' But, on the other hand, the author of the great work on *Prognostics* advises us that when the physician is called in he must seek to ascertain the nature of the affections that he is treating, and especially 'if there be anything divine in the disease, and to learn a foreknowledge of this also.'[113] We may note too that this sentence almost immediately precedes what is perhaps the most famous of all the Hippocratic sentences, the description of what has since been termed the *Hippocratic facies*. This wonderful description of the signs of death may be given as an illustration of the habitual attitude of the Hippocratic school towards prognosis and of the very careful way in which they noted details:

> 'He [the physician] should observe thus in acute diseases: first, the countenance of the patient, if it be like to those who are in health, *and especially if it be like itself, for this would be the best*; but the more unlike to this, the worse it is; such would be

these: *sharp nose, hollow eyes, collapsed temples*; *ears cold, contracted, and their lobes turned out*; *skin about the forehead rough, distended, and parched*; *the colour of the whole face greenish or dusky*. If the countenance be so at the beginning of the disease, and if this cannot be accounted for from the other symptoms, inquiry must be made whether he has passed a sleepless night; whether his bowels have been very loose; or whether he is suffering from hunger; and if any of these be admitted the danger may be reckoned as less; and it may be judged in the course of a day and night if the appearance of the countenance proceed from these. But if none of these be said to exist, and the symptoms do not subside in that time, be it known for certain that death is at hand.'[114]

Again, in the work *On the Art [of Medicine]* we read: 'I hold it to be physicianly to abstain from treating those who are overwhelmed by disease',[115] a prudent if inhumane procedure among a people who might regard the doctor's powers as partaking of the nature of magic, and perhaps a wise course to follow at this day in some places not very far from Cos. Yet in the book *On Diseases* we are advised even in the presence of an incurable disease 'to give relief with such treatment as is possible'.[116]

Furthermore, works by authors of the Hippocratic school stand sometimes in a position of direct controversy with each other. Thus in the treatise *On the Heart* an experiment is set forth which is held to prove that a part at least of imbibed fluid passes into the cavity of the lung and thence to the parts of the body, a popular error in antiquity which recurs in Plato's *Timaeus*. This view, however, is specifically held to be fallacious by the author of the work *On Diseases*, who is supported by a polemical section in the surviving Menon fragment.

Passages like these have convinced all students that we have to deal in this collection with a variety of works written at different dates by different authors and under different conditions, a state that may be well understood when we reflect that among the Greeks medicine was a progressive study for a far longer period of time than has yet been the case in the Western world. An account of such a collection can therefore only be given in the most general fashion. The system or systems of medicine that we shall thus attempt to describe was in vogue up to the Alexandrian period, that is, to the beginning of the third century B. C.

Anatomy and physiology, the basis of our modern system, was still a very weak point in the knowledge of the pre-Alexandrians. The surface form of the body was intimately studied in connexion especially with fractures, but there is no evidence in the literature of the period of any closer acquaintance with human anatomical structure.[117] The same fact is well borne out by Greek Art, for in its noblest period the artist betrays no evidence of assistance derived from anatomization. Such evidence is not found until we come to sculpture of Alexandrian date, when the somewhat strained attitudes and exaggerated musculature of certain works of the school of Pergamon suggest that the artist derived hints, if not direct information, from anatomists who, we know, were active at that time. It is not improbable, however, that separate bones, if not complete skeletons, were commonly studied earlier, for the surgical works of the Hippocratic collection, and especially those on fractures and dislocations, give evidence of a knowledge of the relations of bones to each other and of their natural position in the body which could not be obtained, or only obtained with greatest difficulty, without this aid.

There are in the Hippocratic works a certain number of comparisons between human and animal structures that would have been made possible by surgical operations and occasional accidents. The view has been put forward that some anatomical knowledge was derived through the practice of augury from the entrails of sacrificial animals. It appears, however, improbable that a system so scientific and so little related to temple practice would have had much to learn from these sources, and, moreover, since we know that animals were actually dissected as early as the time of Alcmaeon it would be unnecessary to invoke the aid of the priests. The unknown author of the περὶ τόπων τῶν κατὰ ἄνθρωπον, *On the sites of [diseases] in man*, a work written about 400 B. C., declares indeed that 'physical structure is the basis of medicine', but the formal treatises on anatomy that we possess from Hippocratic times give the general anatomical standard of the corpus, and it is a very disappointing one. The tract *On Anatomy*, though probably of much later date (perhaps *c.* 330 B. C.), is inferior even to the treatise *On the Heart* (perhaps of about 400 B. C.).

Physiology and Pathology are almost as much in the background as anatomy in the Hippocratic collection. As a formal discipline and part of medical education we find no trace of these studies among the pre-Alexandrian physicians. But the meagreness of the number of ascertained facts did not prevent much speculation among a people eager to seek the causes of things. Of that speculation we learn much from the fragments of contemporary medical writers and philosophers, from the medical works of the Alexandrian period, and to some extent from the Hippocratic writings themselves. But the wiser and more sober among the writers of the Hippocratic corpus were bent on something other than the causes of things. Their pre-occupation was primarily with the suffering patient, and the best of them therefore excluded—and we may assume consciously—all but the rarest references to such speculation.

The general state of health of the body was considered by the Hippocratists to depend on the distribution of the four elements, earth, air, fire, and water, whose mixture (*crasis*) and cardinal properties, dryness, warmth, coldness, and moistness, form the body and its constituents. To these correspond the cardinal fluids, blood, phlegm, yellow bile and black bile. The fundamental condition of life is the *innate heat*, the abdication of which is death. This innate heat is greatest in youth when most fuel is therefore required, but gradually declines with age. Another necessity for the support of life is the *pneuma* which circulates in the vessels. All this may seem fanciful enough, but we may remember that the first half of the nineteenth century had waned before the doctrine of the humours which had then lasted for at least twenty-two centuries became obsolete, and perhaps it still survives in certain modern scientific developments. Moreover, the finest and most characteristic of the Hippocratic works either do not mention or but casually refer to these theories which are not essential to their main pre-occupation. Their task of observation of symptoms, of the separation of the essentials from the accidents of disease, and of generalization from experience could go on unaffected by any view of the nature of man and of the world. Even treatment, which must almost of necessity be based on *some* theory of causation, was little deflected by a view of elements and humours on which it was impossible to act directly, while therapeutics was further safeguarded from such influence by the doctrine of *Nature as the healer of diseases*, νούσων φύσεις ἰητροί, the *vis medicatrix naturae* of the later Latin writers and of the present day.

Diseases are to be cured, in the Hippocratic view, by restoring the disturbed harmony in the relation of the elements and humours. These, in fact, tend naturally to an equilibrium and in most cases if left to themselves will be brought to this state by the natural tendency to recovery. The process is known as *pepsis* or, to give it the Latin form, *coctio*, and the turning-point at which the effects of this process exhibit themselves is the *crisis*, a term which, together with some of its original content, has still a place in medicine. Such a turning-point does in fact occur in many diseases, especially those of a zymotic character, on certain special days, though undue emphasis was laid by the Greek physicians upon the exact numerical character of the event. It was no unimportant duty of the physician to assist nature by bringing his remedies to bear at the critical times. If the crisis is wanting, or if the remedies are applied at the wrong moment, the disease may become incurable. But diseases were only immediately or proximately caused by disturbances in the balance or harmony of the humours. This was a mere hypothesis, as the Hippocratists themselves well knew. There were other more remote causes which came into the actual purview of the physician, conditions which he could and did study. Such conditions were, for instance, injudicious modes of life, exposure to climatic changes, advancing age, and the like. Many of these could be directly corrected. But for those that could not there were various therapeutic measures at hand.

That human bodies are and normally remain in a state of health, and that on the whole they tend to recover from disease, is an attitude so familiar to us to-day that we scarcely need to be reminded of it. We live some twenty-three centuries later than Hippocrates; for some sixteen of those centuries the civilized world thought that to retain health periodical bleedings and potions were necessary; for the last century or two we have been gradually returning on the Hippocratic position!

The chief glory of the Hippocratic collection regarded from the clinical point of view is perhaps the actual description of cases. A number of these—forty-two in all—have survived.[118] They are not only unique as a collection for nearly 2,000 years, but they are still to this day models of what succinct clinical records should be, clear and short, without a superfluous word, yet with all that is most essential, and exhibiting merely a desire to record the most important facts without the least attempt to prejudge the case. They illustrate to the full the Greek genius for seizing on the essential. The writer show's not the least wish to exalt his own skill. He seeks merely to put the data before the reader for his guidance under like circumstances. It is a reflex of the spirit of full honesty in which these men lived and worked that the great majority of the cases are recorded to have died. Two of this remarkable little collection may be given:

> 'The woman with quinsy, who lodged with Ariston: her complaint began in the tongue; voice inarticulate; tongue red and parched. *First day*, shivered, then became heated. *Third day*, rigor, acute fever; reddish and hard swelling on both sides of neck and chest; extremities cold and livid; respiration elevated; drink returned by the nose; she could not swallow; alvine and urinary discharges suppressed. *Fourth day*, all symptoms exacerbated. *Fifth day*, she died.'

We probably have here to do with a case of diphtheria. The quinsy, the paralysis of the palate leading to return of the food through the nose, and the difficulty with speech and swallowing are typical results of this affection which was here complicated by a spread of the septic processes into the neck and chest, a not uncommon sequela of the disease. The rapid onset of the conditions is rather unusual, but may be explained if we regard the case as a mild and unnoticed diphtheria, subsequently complicated by paralysis and by secondary septic infection, for which reasons she came under observation.

> 'In Thasos, the wife of Delearces who lodged on the plain, through sorrow was seized with an acute and shivering fever. From first to last she always wrapped herself up in her bedclothes; kept silent, fumbled, picked, bored and gathered hairs [from the clothes]; tears, and again laughter; no sleep; bowels irritable, but passed nothing; when urged drank a little; urine thin and scanty; to the touch the fever was slight; coldness of the extremities. *Ninth day*, talked much incoherently, and again sank into silence. *Fourteenth day*, breathing rare, large, and spaced, and again hurried. *Seventeenth day*, after stimulation of the bowels she passed even drinks, nor could retain anything; totally insensible; skin parched and tense. *Twentieth day*, much talk, and again became composed, then voiceless; respiration hurried. *Twenty-first day*, died. Her respiration throughout was rare and large; she

was totally insensible; always
wrapped up in her bedclothes;
throughout either much talk, or
complete silence.'

This second case is in part a description of low muttering delirium, a common end of continued fevers such as, for instance, typhoid. The description closely resembles the condition known now in medicine as the 'typhoid state'. Incidentally the case contains a reference to a type of breathing common among the dying. The respiration becomes deep and slow, as it sinks gradually into quietude and becomes rarer and rarer until it seems to cease altogether, and then it gradually becomes more rapid and so on alternately. This type of breathing is known to physicians as 'Cheyne-Stokes' respiration in commemoration of two distinguished Irish physicians of the last century who brought it to the attention of medical men.[119] Recently it has been partially explained on a physiological basis. We may note that there is another and even better pen-picture of Cheyne-Stokes respiration in the Hippocratic collection. It is in the famous case of 'Philescos who lived by the wall and who took to his bed on the first day of acute fever'. About the middle of the sixth day he died and the physician notes that 'the respiration throughout was *like that of a person recollecting himself* and was large and rare'. Cheyne-Stokes breathing is admirably described as 'that of a person recollecting himself'.

Such records as these may be contrasted with certain others that have come down from Greek antiquity. We may instance two steles discovered at Epidaurus in 1885, bearing accounts of forty-four temple cures. The following two are fair samples of the cures there described:

'*Aristagora of Troizen.* She had tape-worm, and while she slept in the Temple of Asclepius at Troizen, she saw a vision. She thought that, as the god was not present, but was away in Epidaurus, his sons cut off her head, but were unable to put it back again. Then they sent a messenger to Asklepius asking him to come to Troizen. Meanwhile day came, and the priest actually saw her head cut off from the body. The next night Aristagora had a dream. She thought the god came from Epidaurus and fastened her head on to her neck. Then he cut open her belly, and stitched it up again. So she was cured.'

'A man had an abdominal abscess. He saw a vision, and thought that the god ordered the slaves who accompanied him to lift him up and hold him, so that his abdomen could be cut open. The man tried to get away, but his slaves caught him and bound him. So Asclepius cut him open, rid him of the abscess, and then stitched him up again, releasing him from his bonds. Straightway he departed cured, and the floor of the Abaton was covered with blood.'[120]

In the records of almost all temple cures, a great number of which have survived in a wide variety of documents, an essential element is the process of ἐγκοίμησις, *incubation* or temple sleep, usually in a special sleeping-place or Abaton. The process has a close parallel in certain modern Greek churches and in places of worship much further West; there are even traces of it in these islands, and it is more than probable that the Christian practice is descended by direct continuity from the pagan.[121] The whole character of the temple treatment was—and is—of a kind to suggest to the patient that he should dream of the god, an event which therefore usually takes place. Such treatment by suggestion is applicable only to certain classes of disease and is always liable to fall into the hands of fanatics and impostors. The difficulty that the honest practitioner encounters is that the sufferer, in the nature of the case, can hardly be brought to believe that his ailment is what in fact it is, a lesion of the mind. It is this which gives the miracle-monger his chance.

Examine for a moment the two cases from Epidaurus, which are quite typical of the series. We observe that the first is described simply as a case of 'tape-worm' without any justification for the diagnosis. It is not unfrequent nowadays for thin and anxious patients to state, similarly without justification, that they suffer from this condition. They attribute certain common gastric experiences to this cause of which perhaps they have learned from sensational advertisements, and then they ask cure for a condition which they themselves have diagnosed, but which has no existence in fact. Such a case is often appropriately treated by suggestion. Though the elaborateness of the suggestion in the temple cure is a little startling, yet it can easily be paralleled from the legends of the Christian saints. Moreover, we must remember that we are not here dealing with an account set down by the patient herself, but with an edificatory inscription put up by the temple officials.

In the second inscription, the man with an abdominal abscess, we have a much simpler state of affairs. It is evident that an operation was actually performed by the priest masquerading as Asclepius, while the patient was held down by the slaves. He is assured that all is a dream and departs cured with the tell-tale comment 'and the floor of the Abaton was covered with blood'.

These cases might be multiplied indefinitely without great profit for our particular theme, for in such matters there is no development, no evolution, no history. There can be no doubt that a very large part of Greek practice was on this level, as is a small part of modern medicine, but it is not a level with which we are here dealing and we shall therefore pass it by. But a word of caution must be added. Such temple worship has been compared with modern psycho-analysis. That method, like all methods, has doubtless been abused at times; but it is in essence, unlike the temple system, a purely scientific process by which the ultimate basis of the patient's delusions are laid bare and demonstrated to him.

There is indeed another side to these Asclepian temples. They gradually developed along the lines of our health resorts and developed many of the qualities—lovely and unlovely—that we associate with certain continental watering places. On the bad side they became gossiping centres or even something little better than brothels, as we may gather from the *Mimes* of Herondas. On the good side they formed a quiet refuge among beautiful and interesting surroundings where the sick, exhausted, and convalescent might gain the benefits that accrue from pure air, fine scenery, and a regular and regulated mode of life. It is more than probable too that the open air and manner of living benefited many cases of incipient phthisis.

Returning to the Hippocratic collection, the purely surgical treatises will be found no less remarkable than those of clinical observation. A very able surgeon, Francis Adams (1796-1861), who was eminent as a Greek scholar, gave it as his opinion in the middle of the nineteenth century that no systematic writer on surgery up to his time had given so good and so complete an account of certain dislocations, notably of the hip-joint, as that to be found in the Hippocratic collection. Some types of injury to the hip, as described in the Hippocratic writings, were certainly otherwise quite inadequately known until described by Sir Astley Cooper (1768-1841), himself a peculiarly Hippocratic character.[122] The verdict of Adams was probably just, though since his time the surgery of dislocations, aided especially by X-rays, has been enabled to pass very definitely beyond the Hippocratic position. Admirable, too, is the Hippocratic description of dislocation of the shoulder and of the jaw. In dislocation of hip, shoulder, or jaw, as in most similar lesions, there is considerable deformity produced. The nature and meaning of this deformity is described with remarkable exactness by the Hippocratic writer, who also sets forth the resulting disability. The principles and indeed the very details of treatment in these cases are, save for the use of an anaesthetic, practically identical with those of the present day. The processes are unfortunately not suitable for detailed quotation and description here, but they are of special interest since a graphic record of them has come down to us. There exists in the Laurentian Library at Florence a ninth century Greek surgical manuscript which contains figures of surgeons reducing the dislocations in question. There is good reason to suppose that these miniatures are copied from figures first prepared in pre-Christian times many centuries earlier, and we may here see the actual processes of reduction of such fractures, as conducted by a surgeon of the direct Hippocratic tradition[123] (see Figs. 3, 4).

In keeping with all this is most of the surgical work of the collection. We are almost startled by the modern sound of the whole procedure as we run through the rough notebook κατ᾽ ἰητρεῖον, *Concerning the Surgery*, or the more elaborate treatise περὶ ἰητροῦ, *On the Physician*, where we may read minute directions for the preparation of the operating-room, and on such points as the management of light both artificial and natural, scrupulous cleanliness of the hands, the care and use of the instruments, with the special precautions needed when they are of iron, the decencies to be observed during the operation, the general method of bandaging, the placing of the patient, the use and abuse of splints, and the need for tidiness, order, and cleanliness. Many of these directions are enlarged upon in other surgical works of the collection, among which we find especially full instructions for bandaging and for the diagnosis and treatment of fractures and dislocations. A very fair representation of such a surgery as these works describe is to be found on a vase-painting of Ionic origin which is of the fifth century and therefore about contemporary with

Hippocrates himself (<u>see fig. 5</u>). There are also several beautiful representations on vases of the actual processes of bandaging (<u>fig. 6</u>).

From MS. of APOLLONIUS OF KITIUM, of Ninth
Century
Copied from a pre-Christian original

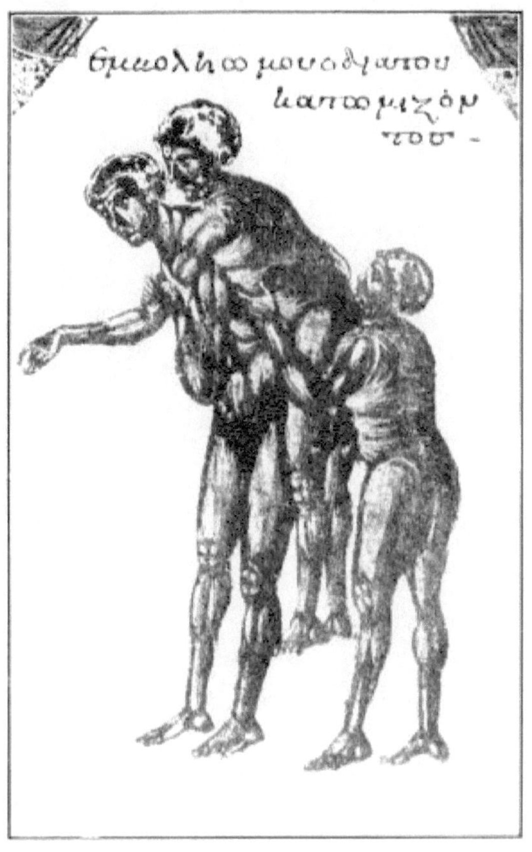

Fig. 3.
REDUCING DISLOCATED SHOULDER

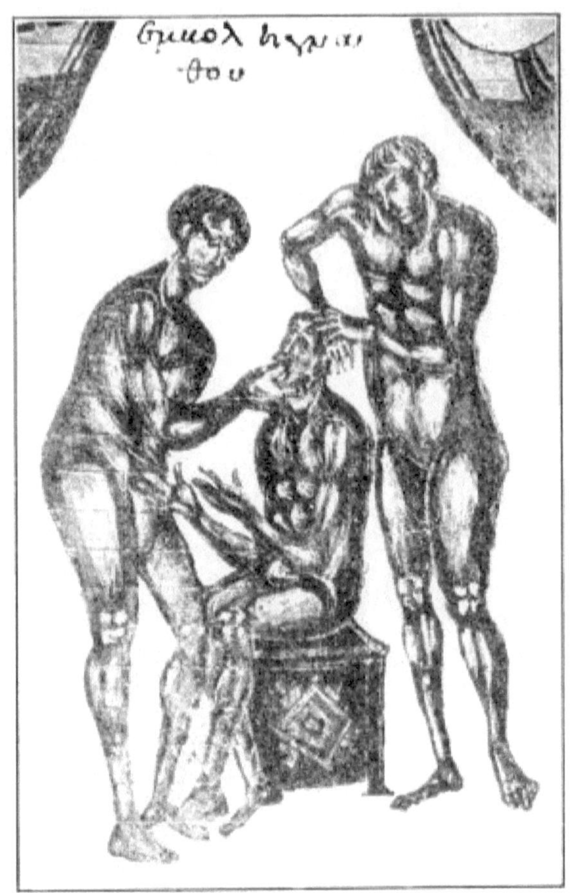

Fig. 4.
REDUCING DISLOCATED JAW

Among the surgical procedures of which descriptions are to be found in the Hippocratic writings are the opening of the chest for the condition known as *empyema* (accumulation of pus within the pleura frequently following pneumonia), and trephining the skull in cases of fracture of that part—two fundamental operations of modern surgery. Surgical art has advanced enormously in our own times, yet a text-book containing much that is useful to this day might be prepared from these surgical contents of the collection alone.

When we pass to the works on Medicine, in the restricted sense, we enter into a region more difficult and perhaps even more fascinating. We are no longer dealing with simple lesions of known origin, but with the effects of disease and degeneration, of the essential character of which the Hippocratic writers could in the nature of the case know very little. Rigidly guarding themselves from any attempt to explain disease by more immediate and hypothetical causes and thus diverting the reader's energies in the medically useless direction of vague speculation—the prevalent mental vice of the Greeks—the best of these physicians are content if they can put forward generalized conclusions from actually observed cases. Many of their thoughts have now become household words, and they have become so, largely as a direct heritage from these ancient physicians. But it must be remembered that ideas so familiar to us were with them the result of long and carefully recorded experience and are like nothing that we encounter in the medicine of other ancient nations. Such conclusions are best set forth perhaps in the wonderful book of the *Aphorisms* from which we may permit ourselves a few quotations:

FIG. 5. A GREEK CLINIC OF ABOUT 400 B.C.
From a vase-painting.

In the centre sits a physician holding a lancet and bleeding a patient from the median vein at the bend of the right elbow into a large open basin. Above and behind the physician are suspended three cupping vessels. To the right sits another patient awaiting his turn; his left arm is bandaged in the region of the biceps. The figure beyond him smells a flower, perhaps as a preservative against infection. Behind the physician stands a man leaning on a staff; he is wounded in the left leg, which is bandaged. By his side stands a dwarfish figure with disproportionately large head, whose body exhibits deformities typical of the developmental disease now known as *Achondroplasia*; in addition to these deformities we note that his body is hairy and the bridge of his nose sunken; on his back he carries a hare which is almost as tall as himself. Talking to the dwarf is a man leaning on a long staff, who has the remains of a bandage round his chest.

See E. Pottier, 'Une Clinique grecque au Ve siècle (vase antique du collection Peztel)', *Fondation Eugène Piot, Monuments et Mémoires*, xiii. 149, Paris, 1906. (Some of our interpretations differ from those of M. Pottier.)

FIG. 6.

A kylix from the Berlin Museum of about 490 B. C. It bears the inscription **ΣΟΣΙΑΣ ΕΠΟΙΗΣΕΝ**, *Sosias made (me)*, and represents Achilles bandaging Patroclus, the names of the two heroes being written round the margin. The painter is Euphronios, and the work is regarded as the masterpiece of that great artist. The left upper arm of Patroclus is injured, and Achilles is bandaging it with a two-rolled bandage, which he is trying to bring down to extend over the elbow. The treatment of the hands, a department in which Euphronios excelled, is particularly fine. Achilles was not a trained surgeon, and it will be observed, from the position of the two tails of the

bandage, that he will have some difficulty
when it comes to its final fastening!

'Life is short, and the Art long; the opportunity fleeting; experiment dangerous, and judgement difficult. Yet we must be prepared not only to do our duty ourselves, but also patient, attendants, and external circumstances must co-operate.'[124]

In this one memorable paragraph, so condensed in the original as to be almost untranslatable, he who 'first separated medicine from philosophy' puts aside at once all speculative interest while in the actual presence of the sick. His whole energy is concentrated on the case in hand with that peculiar attitude, at once impersonal and intensely personal, that has since been the mark of the physician, and that has made of Medicine both a science and an art.

'For extreme diseases, extreme methods of cure.'[125]

'The aged endure fasting most easily; next adults; next young persons, and least of all children, and especially such as are the most lively.'

'Growing bodies have the most innate heat; they therefore require the most nourishment, and if they have it not they waste. In the aged there is little heat, and therefore they require little fuel, for it would be extinguished by much. Similarly fevers in the aged are not so acute, because their bodies are cold.'

'In disease sleep that is laborious is a deadly symptom; but if sleep relieves it is not deadly.'

'Sleep that puts an end to delirium is a good symptom.'

'If a convalescent eats well, but does not put on flesh, it is a bad symptom.'

'Food or drink which is a little less good but more palatable, is to be preferred to such that is better but less palatable.'

'The old have generally fewer complaints than young; but those chronic diseases which do befall them generally never leave them.'

Here we have a group of observations, some of which have become literally household words, nor is it difficult to understand how such sayings have passed from professional into lay keeping. This magnificent book of *Aphorisms* was very early translated into Latin, probably before and certainly not later than the sixth century of the Christian era, and thus became accessible throughout the West. Manuscripts of this Latin version, dating from the ninth and tenth centuries of our era, have survived in the actual places in which they were written, at Monte Cassino in Southern Italy and at Einsiedeln in Switzerland, and in 991 the book of *Aphorisms* was well known and closely studied at the Cathedral school of Chartres. From France the *Aphorisms* reached England, and they are mentioned in documents of the tenth or eleventh century. By now, too, the book had been translated into Syriac and later into Arabic and Hebrew, so that in the true mediaeval period it was known both East and West, and in the vernacular as well as the classical tongues. From the oriental dialects several further translations were again made into Latin. An enormous number of manuscripts of the work have survived in almost every Western dialect, and these show on the whole that the text has been surprisingly little tampered with. In the middle of the thirteenth century some of the better-known Aphorisms were absorbed into a very popular Latin poem that went forth in the name of the medical school of Salerno, though with a false ascription to a yet earlier date. The Salernitan poem, being itself translated into every European vernacular, further helped to bring Hippocrates into every home.

But by no means all the Aphorisms are of a kind that could well become absorbed into folk medicine. It is only those concerning frequently recurring states to which this fate could befall. The book contains also a number of notes on rare conditions seldom seen or noted save by medical men. Such are the following very acute observations:

'Spasm supervening on a wound is fatal.'

'Those seized with tetanus die within four days, or if they survive so long they recover.'

'A convulsion, or hiccup, supervening on a copious discharge of blood is bad.'

'If after severe and grave wounds no swelling appears, it is very serious.'

These four sentences all concern wounds. The first two refer to the disease *tetanus*, which is very liable to supervene on wounds fouled with earth, especially in hot and moist localities. The disease is characterized by a series of painful muscular contractions which in the more severe and fatal form may become a continuous spasm, a type that is referred to in the first sentence. It is true of tetanus that the later the onset after the wound is sustained the better the chance of recovery. This is brought out by the second sentence. The third and fourth sentences record untoward symptoms following a severe wound, now well recognized and watched for by every surgeon. There were, of course, innumerable illustrations of the truth of these Aphorisms in extensive wounds, especially those involving crushed limbs, in the late war.

'Phthisis occurs most commonly between the ages of eighteen and thirty-five.'

'Diarrhœa supervening on phthisis is mortal.'

The period given by the *Aphorisms* for the maximum frequency of onset of the disease is closely borne out by modern observations. The second Aphorism is equally valid; continued diarrhœa is a very frequent antecedent of the fatal event in chronic phthisis, and post-mortem examination has shown that secondary involvement of the bowel is an exceedingly common condition in this disease.

No less remarkable is the following saying: 'In jaundice it is a grave matter if the liver becomes indurated.' Jaundice is a common and comparatively trivial symptom following or accompanying a large variety of diseases. In and by itself it is of little importance and almost always disappears spontaneously. There is a small group of pathological conditions, however, in which this is not the case. The commonest and most important of these are the fatal affections of cirrhosis and cancer of the liver in which that organ may be felt to be enlarged and hardened. If therefore the liver can be so felt in a case of jaundice, it is, as the Aphorism says, of gravest import. Representations of such cases have actually come down to us from Greek times. Thus on a monument erected at Athens to the memory of a physician who died in the second century of the Christian era we may see the process of clinical examination (fig. 7). The physician is palpating the liver of a dwarfish figure whose swollen belly, wasted limbs, and anxious look tell of some such condition as that described in the Aphorism. The ridge caused by the enlarged liver can even be detected on the statue.

'We must attend to the appearances of the eyes in sleep as presented from below; for if a portion of the white be seen between the closing eyelids, and if this be not connected with diarrhœa or severe purging, it is a very bad and mortal symptom.' In this, the last Aphorism which we shall quote, we see the Hippocratic physician actually making his observations. Now during sleep the eyeball is turned upward, so that if the eye be then opened and examined only the white is seen. In the later stages of all wasting and chronic diseases the eyelids tend not to be closed during sleep. Such patients, as is well known, often die with the eyes open and sometimes exhibiting only the whites.

But the Hippocratic physician was not content to make only passive observation; he also took active measures to elicit the 'physical signs'. In modern times a large, perhaps the chief, task of the student of medicine is to acquire a knowledge of these so-called physical signs of disease, the tradition of which has been gradually rebuilt during the last three centuries. Among the most important measures in which he learns to acquire facility is that of auscultation. This useful process has come specially into vogue since the invention of the stethoscope in 1819 by Laennec, who derived valuable hints for it from the Hippocratic writings. Auscultation is several times mentioned and described by the Hippocratic physicians, who used the direct method of listening and not the mediate method devised by Laennec. There are, however, certain cases in which the modern physician still finds the older non-instrumental Hippocratic method superior. In the Hippocratic work περὶ νούσων, *On diseases*, we read of a case with fluid in the pleura that 'you will place the patient on a seat which does not move, an assistant will hold him by the shoulders, and you will shake him, applying the ear to the chest, so as to recognize on which side the sign occurs'. This sign is still used by physicians and is known as *Hippocratic succussion*. In another passage in the same work the symptoms of pleurisy are described and 'a creak like that of leather may be heard'. This is the well known *pleuritic rub* which the physician is accustomed to seek in such cases, and of which the creak of leather is an excellent representation.

Such quotations give an insight into the general method and attitude of the Hippocratics. Of an art such as medicine, which even in those times had a long and rational tradition behind it, it is impossible to give more than the merest glimpse in such a review as this. The actual practice is far too complex to set down briefly. This is especially the case with the ancient teaching as regards epidemic disease at which we must cursorily glance. The Hippocratic physicians and indeed all antiquity were as yet ignorant of the nature, and were but dimly aware of the existence, of infection.[126] For them acute disease was something imposed on the patient from outside, but how it reached him from outside and what it was that thus reached him they were still admittedly ignorant. In this dilemma they turned to prolonged observation and noted as a result of repeated experience that epidemic diseases in their world had characteristic seasonal and regional distributions. One country was not quite like another, nor was one season like another nor even one year like another. By a series of carefully collated observations as to how regions, seasons, and years differed from each other, they succeeded in laying the basis of a rational study of epidemiology which gave rise to the notion of an 'epidemic constitution' of the different years, a conception which was very fertile and stimulating to the great clinicians of the seventeenth and eighteenth centuries and is by no means without value even for the modern epidemiologist. The work of the modern fathers of epidemiology was consciously based on Hippocrates.

Before parting with the Hippocratic physician a word must be said as to his therapeutic means. His general armoury may be described as resembling that of the modern physician of about two generations ago. During those two generations we have, it is true, added to our list of effective remedies but, on the other hand, there has been by common consent a return to the Hippocratic simplicity of treatment. After rest and quiet the central factor in treatment was

Dietetics. This science regarded the age—'Old persons use less nutriment than young'; the season—'In winter abundant nourishment is wholesome, in summer a more frugal diet'; the bodily condition—'Lean persons should take little food, but this little should be fat, fat persons on the other hand should take much food, but it should be lean'. Respect was also paid to the digestibility of different foods—'white meat is more easily digestible than dark'—and to their preparation. Water, barley water, and lime water were recommended as drinks. The dietetic principles of the Hippocratics, especially in connexion with fevers, are substantially those of the present day, and it may be said that the general medical tendency of the last generation in these matters has been an even closer approximation to the Hippocratic. 'The more we nourish unhealthy bodies the more we injure them'; 'The sick upon whom fever seizes with the greatest severity from the very outset, must at once subject themselves to a rigid diet'; 'Complete abstinence often acts well, if the strength of the patient can in any way sustain it'; yet 'We should examine the strength of the sick, to see whether they be in condition to maintain this spare diet to the crisis of the disease'. 'In the application of these rules we must always be mindful of the strength of the patient and of the course of each particular disease, as well as of the constitution and ordinary mode of life in each disease.'

Besides diet the Hippocratic physician had at his disposal a considerable variety of other remedies. Baths, inunctions, clysters, warm and cold suffusions, massage and gymnastic, as well as gentler exercise are among them. He probably employed cupping and bleeding rather too freely, and we have several representations of the instruments used for these operations (fig. 8). He was no great user of drugs and seldom names them except, we may note, in the works on the treatment of women, which are probably of Cnidian origin and whence the greater part of the 300 constituents of

the Hippocratic pharmacopœia are derived. Thus his list of
drugs is small, but several known to him are still used by us.

Fig. 7.   ATHENIAN FUNERARY MONUMENT
Second century A. D. British Museum

Inscription reads: 'Jason, also called Dekmos, the
Acharnian, a physician',
followed by his genealogy. By side of patient stands a
cupping vessel.

The work of these men may be summed up by saying that without dissection, without any experimental physiology or pathology, and without any instrumental aid they pushed the knowledge of the course and origin of disease as far as it is conceivable that men in such circumstances could push it. This was done as a process of pure scientific induction. Their surgery, though hardly based on anatomy, was grounded on the most carefully recorded experience. In therapeutics they allowed themselves neither to be deceived by false hopes nor led aside by vain traditions. Yet in diagnosis, prognosis, surgery and therapeutics alike they were in many departments unsurpassed until the nineteenth century, and to some of their methods we have reverted in the twentieth. Persisting throughout the ages as a more or less definite tradition, which attained clearer form during and after the sixteenth century, Hippocratic methods have formed the basis of all departments of modern advance.

But the history of Greek medicine did not end with the Hippocratic collection; in many respects it may indeed be held only to begin there; yet we never get again a glimpse of so high an ethical and professional standard as that which these works convey. From Alexandrian times onwards, too, the history of Greek medicine becomes largely a history of various schools of medical thought, each of which has only a partial view of the course and nature of medical knowledge. The unravelling of the course and teachings of these sects has long been a pre-occupation of professed medical historians, but the general reader can hardly take an interest in differences between the Dogmatists, Empirics, and Methodists whose doctrines are as dead as themselves. In this later Alexandrian and Hellenistic age the Greek intellect is no less active than before, but there is a change in the taste of the material. A general decay of the spirit is reflected in the medical as in the literary products of the time, and we never again feel that elevation of a beautiful and calmly righteous presence that breathes through the Hippocratic collection and gives it a peculiar aroma.

We shall pass over the general course of later Greek medicine with great rapidity. A definite medical school was established at Alexandria and others perhaps at Pergamon and elsewhere. Athens, after the death of Aristotle and his pupils, passes entirely into the background and is of no importance so far as medicine is concerned. At Alexandria, where a great medical library was collected, anatomy began to be studied and two men whose discoveries were of primary importance for the history of that subject, Erasistratus and Herophilus, early practised there. With anatomy as a basis medical education could become much more systematic. It is a very great misfortune that the works of these two eminent men have disappeared. Of Herophilus fragments have survived embedded in the works of Galen (A. D. 130-201), Caelius Aurelianus (fifth century), and others. These fragments have been the subject of one of the earliest, most laborious, and most successful attempts made in modern times to reconstruct the lost work of an ancient author.[127] For Erasistratus our chief source of information are two polemical treatises directed against him by Galen. Recently, too, a little more information concerning the works of both men has become available from the Menon papyrus.

It has been found possible to reconstruct especially a treatise on anatomy by Herophilus with a considerable show of probability. He opened by giving general directions for the process of dissection and followed with detailed descriptions of the various systems, nervous, vascular, glandular, digestive, generative, and osseous. There was a separate section on the liver, a small part of which has survived. It is of his account of the nervous system that we have perhaps the best record, and it is evident that he has advanced far beyond the Hippocratic position. In the braincase he saw the membranes that cover the brain and distinguished between the cerebrum and cerebellum. He attained to some knowledge of the ventricles of the brain, the cranial and spinal nerves, the nerves of the heart, and the coats of the eye. He distinguished the blood sinuses of the skull, and the *torcular Herophili* (winepress of Herophilus), a sinus described by him, has preserved his name in modern anatomical nomenclature. He even made out more minute structures, such as the little depression in the fourth ventricle of the brain, known to modern anatomists as the *calamus scriptorius*, which still bears the name which he gave it (κάλαμος ᾧ γράφομεν), because it seemed to him, as Galen tells us, to resemble the pens then in use in Alexandria.[128] We still use, too, his term *duodenum* (δωδεκαδάκτυλος ἔκφυσις = twelve-finger extension), for as Galen assures us, Herophilus 'so named the first part of the intestine before it is rolled into folds'.[129] The duodenum is a U-shaped section of the intestine following immediately on the stomach. Being fixed down behind the abdominal cavity it cannot be further convoluted, and this accounts for Galen's description of it. It is about twelve fingers' breadth long in the animals dissected by Herophilus.

Erasistratus, the slightly younger Alexandrian contemporary of Herophilus, has the credit of further anatomical discoveries. He described correctly the action of the epiglottis in preventing the entrance of food and drink into the windpipe during the act of swallowing, he saw the lacteal vessels in the mesentery, and pursued further the anatomy of the brain. He improved on the anatomy of the heart, and described the auriculo-ventricular valves and their mode of closure. He distinguished clearly the motor and sensory nerves. He seems to have adopted a definitely experimental attitude—a very rare thing among ancient physicians—and a description of an experiment made by him has recently been recovered. 'If ', he says, 'you take an animal, a bird, for example, and keep it for a time in a jar without giving it food and then weigh it together with its excreta you will find that there is a considerable loss of weight.'[130] The experiment is a simple one, but it was about nineteen hundred years before a modern professor, Sanctorio Santorio (1561-1636), thought of repeating it.[131]

The anatomical advances made by the Alexandrian school naturally reacted on surgical efficiency. The improvement so effected may be gathered, for instance, from an account of the anatomical relationships in certain cases of dislocation of the hip given by the Alexandrian surgeon Hegetor, who lived about 100 B. C. In his book περὶ αἰτιῶν, *On causes [of disease]*, he asks 'why (certain surgeons) do not seek another way of reducing a luxation of the hip.... If the joints of the jaw, shoulder, elbow, knee, finger, &c., can be replaced, the same, they think, must be true of all parts, nor can they give an account of why the femur cannot be put back into its place.... They might have known, however, that from the head of the femur arises a ligament which is inserted into the socket of the hip bone ... and if this ligament is once ruptured the thigh bone cannot be retained in place'.[132] This passage contains the first description of the structure known to modern anatomists as the *ligamentum teres*, a strong fibrous band which unites the head of the femur with the socket into which it fits in the hip bone, like the string that binds the cup and ball of a child's toy. This ligament is ruptured in certain severe cases of dislocation of the hip.

After the establishment of the school at Alexandria, medical teaching rapidly became organized, but throughout the whole course of antiquity it suffered from the absence of anything in the nature of a state diploma. Any one could practise, with the result that many quacks, cranks, and fanatics were to be found among the ranks of the practitioners who often were or had been slaves. The great Alexandrian school, however, did much to preserve some sort of professional standard, and above all its anatomical discipline helped to this end.

Between the founding of the Alexandrian school and Galen we are not rich in medical writings. Apart from fragments and minor productions, the works of only five authors have survived from this period of over four hundred years, namely, Celsus, Dioscorides, Aretaeus of Cappadocia, and two Ephesian authors bearing the names of Rufus and Soranus.

The work of Celsus of the end of the first century B. C. is a Latin treatise, probably translated from Greek, and is the surviving medical volume of a complete cyclopaedia of knowledge. In spite of its unpromising origin it is an excellent compendium of its subject and shows a good deal of advance in many respects beyond the Hippocratic position. The moral tone too is very high, though without the lofty and detached beauty of Hippocrates. Anatomy has greatly improved, and with it surgical procedure, and the work is probably representative of the best Alexandrian practice. The pharmacopœia is more copious, but has not yet become burdensome. The general line of treatment is sensible and humane and the language concise and clear. Among other items he describes dental practice, with the indications for and methods of tooth extraction, the wiring of teeth, and perhaps a dental mirror. There is an excellent account of what might be thought to be the modern operation for removal of the tonsils. Celsus is still commemorated in modern medicine by the *area Celsi*, a not uncommon disease of the skin. The *De re medica* is in fact one of the very best medical text-books that have come down to us from antiquity. It has had a romantic history. Forgotten during the Middle Ages, it was brought to light by the classical scholar Guarino of Verona (1374-1460) in 1426, and a better copy was discovered by his friend Lamola in 1427. Another copy was found by Thomas Parentucelli (1397-1455), afterwards

Pope Nicholas V in 1443, and the text was later studied by Politian (1454-94). Though one of the latest of the great classical medical texts to be discovered, it was one of the first to be printed (Florence, 1478), and it ran through very many early editions and had great influence on the medical renaissance.

After Celsus comes Dioscorides in the first century A. D. He was a Greek military surgeon of Cilician origin who served under Nero, and in him the Greek intellect is obviously beginning to flag. His work is prodigiously important for the history of botany, yet so far as rational medicine is concerned he is almost negligible. He begins at the wrong end, either giving lists of drugs with the symptoms that they are said to cure or to relieve, or lists of symptoms with a series of named drugs. Clinical observation and record are wholly absent, and the spirit of Hippocrates has departed from this elaborate pharmacopœia.

Fig. 8.　VOTIVE TABLET representing cupping and bleeding instruments from Temple of Asclepius at Athens.

In centre is represented a folding case containing scalpels of various forms. On either side are cupping vessels.

With the second century of the Christian era we terminate the creative period of Greek medicine. We are provided with the works of four important writers of this century, of whom three, Rufus of Ephesus, Soranus of Ephesus, and Aretaeus of Cappadocia, though valuable for forming a picture of the state of medicine in their day, were without substantial influence on the course of medicine in later ages.

Rufus of Ephesus, a little junior to Dioscorides, has left us the first formal work on human anatomy and is of some importance in the history of comparative anatomy. In medicine he is memorable as the first to have described bubonic plague, and in surgery for his description of the methods of arresting haemorrhage and his knowledge of the anatomy of the eye. A work by him *On gout* was translated into Latin in the sixth century, but remained unknown till modern times.

Soranus of Ephesus (A. D. *c.* 90-*c.* 150), an acute writer on gynaecology, has left a book which illustrates well the anatomy of his day. It exercised an influence for many centuries to come, and a Latin abstract of it prepared about the sixth century by one Moschion has come down to us in an almost contemporary manuscript.[133] It is interesting as opposing the Hippocratic theory that the male embryo is originated in the right and the female in the left half of the womb, a fallacy derived originally from Empedocles and Parmenides, but perpetuated by Latin translations of the Hippocratic treatises until the seventeenth century. His work was adorned by figures, and some of these, naturally greatly altered by copyists, but still not infinitely removed from the facts, have survived in a manuscript of the ninth century, and give us a distant idea of the appearance of ancient anatomical drawings.[134] We may assist our imagination a little further, in forming an idea of what such diagrams were like, with the help of certain other mediaeval figures representing the form and distribution of the various anatomical 'systems', veins, arteries, nerves, bones, and muscles which are probably traceable to an Alexandrian origin.[135]

Aretaeus of Cappadocia was probably a contemporary of Galen (second half of the second century A. D.). As a clinical author his reputation stands high, perhaps too high, his descriptions of pneumonia, emphysema, diabetes, and elephantiasis having especially drawn attention. In treatment he uses simple remedies, is not affected by polypharmacy, and suggests many ingenious mechanical devices. It would appear that Aretaeus is not an independent writer, but mainly a compiler. He relies largely on Archigenes, a distinguished physician contemporary with Juvenal, whose works have perished save the fragments preserved in this manner by Aretaeus and Aetius. Aretaeus was a very popular writer among the Greeks in all ages, but he was not translated into Latin, and was unknown in the West until the middle of the sixteenth century.[136] He is philologically interesting as still using the Ionic dialect.

There remains the huge overshadowing figure of Galen. The enormous mass of the surviving work of this man, the dictator of medicine until the revival of learning and beyond, tends to throw out of perspective the whole of Greek medical records. The works of Galen alone form about half of the mass of surviving Greek medical writings, and occupy, in the standard edition, twenty-two thick, closely-printed volumes. These cover every department of medicine, anatomy, physiology, pathology, medical theory, therapeutics, as well as clinical medicine and surgery. In style they are verbose and heavy and very frequently polemical. They are saturated with a teleology which, at times, becomes excessively tedious. In the anatomical works, masses of teleological explanation dilute the account of often imperfectly described structures. Yet to this element we owe the preservation of the mass of Galen's works, for his intensely teleological point of view appealed to the theological bias both of Western Christianity and of Eastern Islam. Intolerable as literature, his works are a valuable treasure house of medical knowledge and experience, custom, tradition, and history.

As in the case of the Hippocratic corpus, so in the case of the Galenic corpus we are dealing to some extent with material from various sources. In the case of Galen, however, we have a good standard of genuineness, for he has left us a list of his books which can be checked off against those which we actually possess. The general standpoint of the Galenic is not unlike that of the Hippocratic writings, but the noble vision of the lofty-minded, pure-souled physician has utterly passed away. In his place we have an acute, honest, very contentious fellow, bristling with energy and of prodigious industry, not unkindly, but loving strife, a thoroughly 'aggressive' character. He loves truth, but he loves argument quite as much. The value of his philosophical writings, of which some have survived, cannot be discussed here, but it is evident that he is frequently satisfied with purely verbal explanations. An ingenious physiologist, a born experimenter, an excellent anatomist and eager to improve, possessing a good knowledge of the human skeleton and an accurate acquaintance with the internal parts so far as this can be derived from a most industrious devotion to dissection of animals, equipped with all the learning of the schools of Pergamon, Smyrna, and Alexandria, and rich with the experience of a vast practice at Rome, Galen is essentially an 'efficient' man. He has the grace to acknowledge constantly and repeatedly his indebtedness to the Hippocratic writings. Such was the man whose remains, along with the Hippocratic collection, formed the main medical legacy of Greece to the Western world.

Some of Galen's works are mere drug lists, little superior to those of Dioscorides;[137] with the depression of the intelligence that corresponded with the break up of the Roman Empire, it was these that were chiefly seized on and distributed in the West. Attractive too to the debased intellect of the late Roman world were certain spurious, superstitious, and astrological works that circulated in the name of Galen and Hippocrates.[138] The Greek medical writers after Galen were but his imitators and abstractors, but through some of them Galen's works reached the West at a very early period in the Middle Ages. Such abstractors who were early translated into Latin were Oribasius (325-403), Paul of Aegina (625-690), and Alexander of Tralles (525-605). Of the best and most scientific of Galen's works the Middle Ages knew little or nothing.

Later Galen and Hippocrates became a little more accessible, not by translation from the Greek, but by translation from the Arabic of a Syriac version. The first work to be so rendered was a version of *Aphorisms* of Hippocrates which, however, as we have seen, were already available in Latin dress, together with the Hippocratic *Regimen in acute diseases*, and certain works of Galen as corruptly interpreted by Isaac Judaeus. These were rendered from Arabic into Latin by Constantine, an African adventurer who became a monk at Monte Cassino and died there in 1087. Constantine was a wretched craftsman with an imperfect knowledge of both Arabic and Latin. More effective was the great twelfth century translator from the Arabic, Gerard of Cremona (died 1185), who turned many medical works into Latin from Arabic, and who was followed by a whole host of imitators. Yet more important for the advance of medicine, however, was the learned revival of the thirteenth century. In the main that revival was based on translations from Arabic, but a certain number of works were also rendered direct from the Greek. During the thirteenth century Aristotle's scientific works began to be treated in this way, but more important for the course of medicine were those of Galen, and they had to wait till the following century. The long treatise of Galen, περὶ χρείας τῶν ἐν ἀνθρώπου σώματι μορίων, *On the uses of the bodily parts in man*, was translated from the Greek into Latin by Nicholas of Reggio in the earlier part of the fourteenth century. This work, with all its defects, was by far the best account of the human body then available. Many manuscripts of the Latin version have survived, and it was translated into several vernaculars, including English, and profoundly influenced surgery. The rendering into Latin of this treatise, and its wide distribution, may be regarded as the starting-point of modern scientific medicine. Its appearance is moreover a part of the phenomenon of the revived interest in dissection which had begun to be practised in the Universities in the thirteenth century,[139] and was a generally accepted discipline in the fourteenth and fifteenth.[140]

Until the end of the fifteenth century progress in anatomy was almost imperceptible. During the fifteenth century more Galenic and Hippocratic texts were recovered and gradually turned into Latin, but still without vitally affecting the course of Anatomy. The actual printing of collected editions of Hippocrates and Galen came rather late, for the debased taste of the Renaissance physicians continued to prefer Dioscorides and the Arabs, of whom numerous editions appeared, so that medicine made no advance corresponding to the progress of scholarship. The Hippocratic works were first printed in 1525, and an isolated edition of the inferior Galen in 1490, but the real advance in Medicine was not made by direct study of these works. So long as they were treated in the old scholastic spirit such works were of no more value than those of the Arabists or others inherited from the Middle Ages. Even Hippocrates can be spoilt by a commentary, and it was not until the investigator began actually to compare his own observations with those of Hippocrates and Galen that the real value of these works became apparent. The department in which this happened first was Anatomy, and such revolutionaries as Leonardo da Vinci (1452-1518), who never published, and Vesalius (1514-1564), whose great work appeared in 1543, were really basing their work on Galen, though they were much occupied in proving Galen's errors. Antonio Benivieni (died 1502), an eager prophet of the new spirit, revived the Hippocratic tradition by actually collecting notes of a few cases with accompanying records of deaths and post-mortem findings, among which it is interesting to observe a case of appendicitis.[141] His example was occasionally followed during the sixteenth century, as for instance, by the Portuguese Jewish physician Amatus Lusitanus (1511-*c.* 1562), who printed no fewer than seven hundred cases; but the real revival of the Hippocratic tradition came in the next century with Sydenham (1624-1689) and Boerhaave (1668-1738), who were consciously working on the Hippocratic basis and endeavouring to extend the Hippocratic experience.

Lastly surgery came to profit by the revival. The greatest of the sixteenth century surgeons, the lovable and loving Ambroise Paré (1510-1590), though he was, as he himself humbly confessed, an ignorant man knowing neither Latin nor Greek, can be shown to have derived much from the works of antiquity, which were circulating in translation in his day and were thus filtering down to the unlearned.

Texts of Hippocrates and of Galen had formed an integral part in the medical instruction of the universities from their commencement in the thirteenth century. The first Greek text of the *Aphorisms* of Hippocrates appeared in 1532, edited by no less a hand than that of François Rabelais.

With the further recovery of the Greek texts and preparation of better translations, these became almost the sole mode of instruction during the fifteenth and sixteenth centuries. The translators became legion and their competence varied. One highly skilled translator, however, is of special interest to English readers. Thomas Linacre (1460?-1524), Physician to Henry VIII, Tutor to the Princess Mary, founder and first president of the College of Physicians, a benefactor of both the ancient Universities and one of the earliest, ablest, most typical, and most exasperating of the English humanists, spent much energy on this work of translation for which his abilities peculiarly fitted him. He was responsible for no less than six important works of Galen, of which one, the *De temperamentis et de inaequali intemperie*, printed at Cambridge in 1521, was among the earliest books impressed in that town and is said to be the first printed in England for which Greek types were used. It has been honoured by reproduction in facsimile in modern times. Such works as these, purely literary efforts, had great vogue for a century and more, and were much in use in the Universities. These humanistic products sometimes produced, among the advocates of the new scientific method, a degree of fury which was only rivalled by that of some of the humanists themselves towards the translators from the Arabic. But these are now dead fires. As the clinical and scientific methods of teaching gained ground, textual studies receded in medical education, as Hippocrates and Galen themselves would have wished them to recede.

The texts of Hippocrates and Galen have now ceased to occupy a place in any medical curriculum. Yet all who know these writings, know too, not only that their spirit is still with us, but that the works themselves form the background of modern practice, and that their very phraseology is still in use at the bedside. Modern medicine may be truly described as in essence a creation of the Greeks. To realize the nature of our medical system, some knowledge of its Greek sources is essential. It would indeed be a bad day for medicine if ever this debt to the Greeks were forgotten, and the loss would be at least as much ethical as intellectual. But there is happily no fear of this, for the figure and spirit of Hippocrates are more real and living to-day than they have been since the great collapse of the Greek scientific intellect in the third and fourth centuries of the Christian era.

Footnotes:

[1] The word *Biology* was introduced by Gottfried Reinhold Treviranus (1776-1837) in his *Biologie oder die Philosophie der lebenden Natur*, 6 vols., Göttingen, 1802-22, and was adopted by J.-B. de Lamarck (1744-1829) in his *Hydrogéologie*, Paris, 1802. It is probable that the first English use of the word in its modern sense is by Sir William Lawrence (1783-1867) in his work *On the Physiology, Zoology, and Natural History of Man*, London, 1819; there are earlier English uses of the word, however, contrasted with *biography*.

[2] The remains of Alcmaeon are given in H. Diels' *Die Fragmente der Vorsokratiker*, Berlin, 1903, p. 103. Alcmaeon is considered in the companion chapter on *Greek Medicine*.

[3] Especially the περὶ γυναικείης φύσιος, *On the nature of woman*, and the περὶ γυναικείων, *On (the diseases of) women*.

[4] περὶ ἑβδομάδων. The Greek text is lost. We have, however, an early and barbarous Latin translation, and there has recently been printed an Arabic commentary. G. Bergsträsser, *Pseudogaleni in Hippocratis de septimanis commentarium ab Hunaino Q. F. arabice versum*, Leipzig, 1914.

[5] περὶ νούσων δ'.

[6] περὶ καρδίης.

[7] Especially in the περὶ γονῆς.

[8] The three works περὶ γονῆς, περὶ φύσιος παιδίον, περὶ νούσων δ', *On generation, on the nature of the embryo, on diseases, book IV*, form really one treatise on generation.

[9] περὶ φύσιος παιδίον, *On the nature of the embryo*, § 13. The same experience is described in the περὶ σαρκῶν, *On the muscles*.

[10] περὶ φύσιος παιδίον, *On the nature of the embryo*, § 29.

[11] περὶ φύσιος παιδίον, *On the nature of the embryo*, § 22.

[12] *Ibid.* § 23.

[13] See a valuable note by D'Arcy W. Thompson prefixed to his translation of the *Historia Animalium*, Oxford, 1910.

[14] Pliny, *Naturalis historia*, viii. 17.

[15] Athenaeus, *Deipnosophistae*, ix. 58.

[16] Aelian, *Variae historiae*, iv. 19.

[17] The statement of the relation of Callisthenes to Aristotle rests on the somewhat unsatisfactory evidence of Simplicius (sixth century) who states that Callisthenes sent Aristotle certain astronomical observations from Babylon. Simplicius, *Commentarii* (Karsten), p. 226.

[18] Plutarch, *Alexander*, lv.

[19] The subject is well discussed by W. Ogle in the introduction to his *Aristotle on the Parts of Animals*, London, 1882.

[20] The problem of genuineness is discussed in detail by R. Shute, *On the history of the process by which the Aristotelian writings arrived at their present form*, Oxford, 1888.

[21] I have somewhat abbreviated this and the previous sentence.

[22] *De partibus animalium*, i. 5; 644^b 21.

[23] *De partibus animalium*, i. 1; 641^b 12.

[24] *Physics*, ii. 8, 3; 198^b 6. This passage is considerably abbreviated and slightly paraphrased.

[25] *De partibus animalium*, i. 1; 641^a 7.

[26] *Historia animalium*, viii. 1; 588^b 4.

[27] *De partibus animalium*, iv. 5; 681^a 15.

[28] *De partibus animalium*, iv. 5; 681^a 36.

[29] *De partibus animalium*, iv. 5; 681^a 10.

[30] *De generatione animalium*, i. 21; 729^a 21.

[31] *De generations animalium*, i. 18; 725^a 22.

[32] *De generatione animalium*, i. 19; 727^a 31.

[33] *De generatione animalium*, i. 22; 730^b 10.

[34] *De generatione animalium*, i. 22; 730^a 34.

[35] *Historia animalium*, vi. 3; 561^a 4.

[36] *Cor primum movens ultimum moriens*. This famous sentence is the sense though not the phrasing of *De generatione animalium*, ii. 1 and 4.

[37] *Historia animalium*, vi. 3; 561^a 18.

[38] *De generatione animalium*, iii. 9; 758^a 37.

[39] *Historia animalium*, i. 5; 489^a 35.

[40] *Historia animalium*, vi. 10; 565^b 2.

[41] The history of this discovery is given by Charles Singer, *Studies in the History and Method of Science*, vol. ii, Oxford, 1921, pp. 32 ff.

[42] Johannes Müller, *Ueber den glatten Hai des Aristoteles*, Berlin, 1842.

[43] *De generatione animalium*, ii. 1; 733^a 6.

[44] *Metaphysics*, i. 4. *De generatione et corruptione*, ii. 1.

[45] *De anima*, ii. 1, ii.

[46] *De anima*, ii. 2, ii; 413^a 22.

[47] The question of Aristotle's meaning in connexion with this topic, of primary importance for all thought, has a vast literature. An authoritative work is R. D. Hicks, *Aristotle, De anima*, Cambridge, 1907.

[48] *De partibus animalium*, i. 4; 644^a 22.

[49] *De partibus animalium*, i. 4; 644^a 27.

[50] *De partibus animalium*, i. 4; 644^a 16.

[51] The classificatory system of Aristotle and its history are discussed in great detail by J. B. Meyer, *Aristoteles' Thierkunde: ein Beitrag zur Geschichte der Zoologie, Physiologie und alten Philosophie*, Berlin, 1855.

[52] The work by which Wotton is known is his *De differentiis animalium*, Paris, 1552.

[53] There is a valuable chapter on the subject of the Aristotelian classificatory system as based on the method of reproduction in W. Ogle, *Aristotle on the Parts of Animals*, London, 1882.

[54] The rediscovery and verification of this and other Aristotelian observations is detailed by C. Singer, 'Greek Biology and the Rise of Modern Biology,' *Studies in the History and Method of Science*, vol. ii, Oxford, 1921.

[55] *Historia animalium*, ii. 17; 507^a 33.

[56] *De partibus animalium*, ii. 17; 507^b 12.

[57] *Historia animalium*, ii. 17; 507^b 12.

[58] *Historia animalium*, v. 6; 541^b 1. The hectocotylization of the cephalopod arm which is here recorded as an element in the reproductive

process of these animals is denied in the *De generatione animalium*, i. 15; 720[b] 32, where we read that 'the insertion of the arm of the male into the funnel of the female ... is only for the sake of attachment, and it is not an organ useful for generation, for it is outside the passage in the male and indeed outside the body of the male altogether.' Yet even here Aristotle knows of the physical relationship of the arm. See note on this point in the translation of the passage by A. Platt, Oxford, 1910.

[59] J. B. Verany, *Mollusques méditerranéens*, Genoa, 1851.

[60] E. Racovitza. *Archives de zoologie experimentale*, Paris, 1894.

[61] The paragraphs concerning the fishing-frog and torpedo are made up of sentences rearranged from the *De partibus animalium*, iv. 13; 696[a] 26, and the *Historia animalium*, ix. 37; 620[b] 15.

[62] *De partibus animalium*, ii. 1; 646[a] 12.

[63] *De partibus animalium*, ii. 10.

[64] It is possible that Theophrastus derived the word pericarp from Aristotle. Cp. *De anima*, ii. 1, 412[b] 2. In the passage τὸ φύλλον περικαρπίου σκέπασμα, τὸ δὲ περικάρπιον καρποῦ, in the *De anima* the word does not, however, seem to have the full technical force that Theophrastus gives to it.

[65] *Historia plantarum*, i. 2, vi.

[66] *Ibid.* i. 1, iv.

[67] *Historia plantarum*, ii. 1, i.

[68] *Historia plantarum*, viii. 1, i.

[69] Nathaniel Highmore, *A History of Generation*, London, 1651.

[70] Marcello Malpighi, *Anatome plantarum*, London, 1675.

[71] Nehemiah Grew, *Anatomy of Vegetables begun*, London, 1672.

[72] Pliny, *Naturalis historia*, xiii. 4.

[73] The curious word ὀλυνθάζειν, here translated *to use the wild fig*, is from ὄλυνθος, a kind of wild fig which seldom ripens. The special meaning here given to the word is explained in another work of Theophrastus, *De causis plantarum*, ii. 9, xv. After describing caprification in figs, he says τὸ δὲ ἐπὶ τῶν φοινίκων συμβαῖνον οὐ ταὐτὸν μέν, ἔχει δέ τινα ὁμοιότητα τούτω δι' ὃ καλοῦσιν ὀλυνθάζειν αὐτούς. 'The same thing is not done with dates, but something analogous to it, whence this is called ὀλυνθάζειν'.

[74] *Historia plantarum*, ii. 8, iv.

[75] Herodotus i. 193.

[76] *Historia plantarum*, ii. 8, i.

[77] *Ibid.* ii. 8, ii.

[78] *Historia plantarum*, ii. 8, iv.

[79] *Ibid.* i. 1, ix.

[80] *Ibid.* iii. 18, x.

[81] *De causis plantarum*, ii. 23.

[82] *Historia plantarum*, i. 13, iii.

[83] See the companion chapter on *Greek Medicine.*

[84] The works of Crateuas have recently been printed by M. Wellmann as an appendix to the text of Dioscorides, *De re medica*, 3 vols., Berlin, 1906-17. The source and fate of his plant drawings are discussed in the same author's *Krateuas*, Berlin, 1897.

[85] The manuscript in question is Med. Graec. 1 at what was the Royal Library at Vienna. It is known as the *Constantinopolitanus*. After the war it was taken to St. Mark's at Venice, but either has been or is about to be restored to Vienna. A facsimile of this grand manuscript was published by Sijthoff, Leyden, 1906.

[86] The lady in question was Juliana Anicia, daughter of Anicius Olybrius, Emperor of the West in 472, and his wife Placidia, daughter of Valentinian III. Juliana was betrothed in 479 by the Eastern Emperor Zeno to Theodoric the Ostrogoth, but was married, probably in 487 when the manuscript was presented to her, to Areobindus, a high military officer under the Byzantine Emperor Anastasius.

[87] The importance of this manuscript as well as the position of Dioscorides as medical botanist is discussed by Charles Singer in an article 'Greek Biology and the Rise of Modern Biology'; *Studies in the History and Method of Science*, vol. ii, Oxford, 1921.

[88] This manuscript is at the University Library at Leyden, where it is numbered Voss Q 9.

[89] A good instance of Galen's teleological point of view is afforded by his classical description of *the hand* in the περὶ χρείας τῶν ἐν ἀνθρώπου σώματι μορίων, *On the uses of the parts of the body of man,*

i. 1. This passage is available in English in a tract by Thomas Bellott, London, 1840.

[90] C. H. Haskins, 'The reception of Arabic science in England,' *English Historical Review*, London, 1915, p. 56.

[91] The latest and best work on the Aristotelian translations of Scott is an inaugural dissertation by A. H. Querfeld, *Michael Scottus und seine Schrift, De secretis naturae*, Leipzig, 1919.

[92] J. G. Schneider, *Aristotelis de animalibus historiae*, Leipzig, 1811, p. cxxvi. L. Dittmeyer, *Guilelmi Moerbekensis translatio commentationis Aristotelicae de generatione animalium*, Dillingen, 1915. L. Dittmeyer, *De animalibus historia*, Leipzig, 1907.

[93] The subject of the Latin translations of Aristotle is traversed by A. and C. Jourdain, *Recherches critiques sur l'âge des traductions latines d'Aristote*, 2nd ed., Paris, 1843; M. Grabmann, *Forschungen über die lateinischen Aristoteles-Übersetzungen des XIII. Jahrhunderts*, Münster i/W., 1916; and F. Wüstenfeld, *Die Übersetzungen arabischer Werke in das Lateinische seit dem XI. Jahrhundert*, Göttingen, 1877.

[94] The enormous *De Animalibus* of Albert of Cologne is now available in an edition by H. Stadler, *Albertus Magnus De Animalibus Libri XXVI nach der cölner Urschrift*, 2 vols., Münster i/W., 1916-21. The quotation is translated from vol. i, pp. 465-6.

[95] Conrad's work is conveniently edited by H. Schultz, *Das Buch der Natur von Conrad von Megenberg, die erste Naturgeschichte in deutscher Sprache, in Neu-Hochdeutsche Sprache bearbeitet*, Greifswald, 1897. Conrad's work is based on that of Thomas of Cantimpré (1201-70).

[96] Hieronimo Fabrizio of Acquapendente, *De formato foetu*, Padua, 1604.

[97] William Harvey, *Exercitationes de generatione animalium*, London, 1651.

[98] Karl Ernst von Baer, *Ueber die Entwickelungsgeschichte der Thiere*, Königsberg, 1828-37.

[99] The works of Herophilus are lost. This fine passage has been preserved for us by Sextus Empiricus, a third century physician, in his πρὸς τοὺς μαθηματικοὺς ἀντιρρητικοί, which is in essence an attack on all positive philosophy. It is an entertaining fact that we should have to go to such a work for remains of the greatest anatomist of antiquity. The passage is in the section directed against ethical writers, xi. 50.

[100] The word φυσικός, though it passed over into Latin (Cicero) with the meaning *naturalist*, acquired the connotation of *sorcerer* among the later Greek writers. Perhaps the word *physicianus* was introduced to make a distinction from the charm-mongering *physicus*. In later Latin *physicus* and *medicus* are almost always interchangeable.

[101] This fragment has been published in vol. iii, part 1, of the *Supplementum Aristotelicum* by H. Diels as *Anonymi Londinensis ex Aristotelis Iatricis Menonis et Aliis Medicis Eclogae*, Berlin, 1893. See also H. Bekh and F. Spät, *Anonymus Londinensis, Auszüge eines Unbekannten aus Aristoteles-Menons Handbuch der Medizin*, Berlin, 1896.

[102] It is tempting, also, to connect the Asclepian snake cult with the prominence of the serpent in Minoan religion.

[103] This word *pronoia*, as Galen explains (εἰς τὸ Ἱπποκράτους προγνωστικόν, K. xviii, B. p. 10), is not used in the philosophic sense, as when we ask whether the universe was made by chance or by *pronoia*, nor is it used quite in the modern sense of *prognosis*, though it includes that too. *Pronoia* in Hippocrates means knowing things about a patient before you are told them. See E. T. Withington, 'Some Greek medical terms with reference to Luke and Liddell and Scott,' *Proceedings of the Royal Society of Medicine* (*Section of the History of Medicine*), xiii, p. 124, London, 1920.

[104] *Prognostics* 1.

[105] There is a discussion of the relation of the Asclepiadae to temple practice in an article by E. T. Withington, 'The Asclepiadae and the Priest of Asclepius,' in *Studies in the History and Method of Science*, edited by Charles Singer, vol. ii, Oxford, 1921.

[106] The works of Anaximenes are lost. This phrase of his, however, is preserved by the later writer Aetios.

[107] For the work of these physicians see especially M.Wellmann, *Fragmentsammlung der griechischen Aerzte*, Bd. I, Berlin, 1901.

[108] Galen, περὶ ἀνατομικῶν ἐγχειρήσεων, *On anatomical preparations*, § 1, K. II, p. 282.

[109] *Historia animalium*, iii. 3, where it is ascribed to Polybus. The same passage is, however, repeated twice in the Hippocratic writings, viz. in the περὶ φύσιος ἀνθρώπου, *On the nature of man*, Littré, vi. 58, and in the περὶ ὀστέων φύσιος, *On the nature of bones*, Littré, ix. 174.

[110] Παραγγελίαι, § 6.

[111] It must, however, be admitted that even in the Hippocratic collection itself are cases of breach of the oath. Such, for instance, is the induction of abortion related in περὶ φύσιος παιδίον, *On the nature of the embryo*. There is evidence, however, that the author of this work was not a medical practitioner.

[112] Rome Urbinas 64, fo. 116.

[113] Kühlewein, i. 79, regards this as an interpolated passage.

[114] Littré, ii. 112; Kühlewein, i. 79. The texts vary: Kühlewein is followed except in the last sentence.

[115] Περὶ τέχνες, § 3.

[116] Περὶ νούσων α', § 6.

[117] A reference to dissection in the περὶ ἄρθρων, *On the joints*, § 1, appears of the present writer to be of Alexandrian date.

[118] They are to be found as an Appendix to Books I and III of the *Epidemics* and embedded in Book III.

[119] John Cheyne (1777-1836) described this type of respiration in the *Dublin Hospital Reports*, 1818, ii, p. 216. An extreme case of this condition had been described by Cheyne's namesake George Cheyne (1671-1743) as the famous 'Case of the Hon. Col. Townshend' in his *English Malady*, London, 1733. William Stokes (1804-78) published his account of Cheyne-Stokes breathing in the *Dublin Quarterly Journal of the Medical Sciences*, 1846, ii, p. 73.

[120] The Epidaurian inscriptions are given by M. Fraenkel in the *Corpus Inscriptionum Graecarum* IV, 951-6, and are discussed by Mary Hamilton (Mrs. Guy Dickins), *Incubation*, St. Andrews, 1906, from whose translation I have quoted. Further inscriptions are given by Cavvadias in the *Archaiologike Ephemeris*, 1918, p. 155 (issued 1921).

[121] We are almost told as much in the apocryphal *Gospel of Nicodemus*, § 1, a work probably composed about the end of the fourth century.

[122] Astley Paston Cooper, *Treatise on Dislocations and Fractures of the Joints*, London, 1822, and *Observations on Fractures of the Neck and the Thighbone*, &c., London, 1823.

[123] This famous manuscript is known as Laurentian, Plutarch 74, 7, and its figures have been reproduced by H. Schöne, *Apollonius von Kitium*, Leipzig, 1896.

[124] The first lines are the source of the famous lines in Goethe's *Faust*:

'Ach Gott! die Kunst ist lang Und kurz ist unser Leben, Mir wird bei meinem kritischen Bestreben Doch oft um Kopf und Busen bang.'

[125] The extreme of treatment refers in the original to the extreme restriction of diet, ἐς ἀκριβείην, but the meaning of the Aphorism has always been taken as more generalized.

[126] The ancients knew almost nothing of infection as *applied specifically* to disease. All early peoples—including Greeks and Romans—believed in the transmission of qualities from object to object. Thus purity and impurity and good and bad luck were infections, and diseases were held to be infections in that sense. But there is little evidence in the belief of the special infectivity of *disease as such* in antiquity. Some few diseases are, however, unequivocally referred to as infectious in a limited number of passages, e. g. ophthalmia, scabies, and phthisis in the περὶ διαφορᾶς πυρετῶν, *On the differentiae of fevers*, K. vii, p. 279. The references to infection in antiquity are detailed by C. and D. Singer, 'The scientific position of Girolamo Fracastoro', *Annals of Medical History*, vol. i, New York, 1917.

[127] K. F. H. Marx, *Herophilus, ein Beitrag zur Geschichte der Medizin*, Karlsruhe, 1838.

[128] Galen, περὶ ἀνατομικῶν ἐγχειρήσεων, *On anatomical preparations*, ix. 5 (last sentence).

[129] Galen, περὶ φλεβῶν καὶ ἀρτηριων ανατομῆς, *On the anatomy of veins and arteries*, i.

[130] The quotation is from chapter xxxiii, line 44 of the *Anonymus Londinensis*. H. Diels, *Anonymus Londinensis* in the *Supplementum Aristotelicum*, vol. iii, pars 1, Berlin, 1893.

[131] Sanctorio Santorio, *Oratio in archilyceo patavino anno 1612 habita; de medicina statica aphorismi*. Venice, 1614.

[132] This is the only passage of Hegetor's writing that has survived. It has been preserved in the work of Apollonius of Citium.

[133] Leyden Voss 4° 9* of the sixth century is a fragment of this work.

[134] V. Rose, *Sorani Ephesii vetus translatio Latina cum additis Graeci textus reliquiis*, Leipzig, 1882; F. Weindler, *Geschichte der gynäkologisch-anatomischen Abbildung*, Dresden, 1908.

[135] The discovery and attribution of these figures is the work of K. Sudhoff. A bibliography of his writings on the subject will be found in a 'Study in Early Renaissance Anatomy' in C. Singer's *Studies in the History and Method of Science*, vol. i, Oxford, 1917.

[136] First Latin edition Venice, 1552; first Greek edition Paris, 1554.

[137] e. g. περὶ κράσεως καὶ δυνάμεως τῶν ἀπάντων φαρμάκων and the φάρμακα.

[138] e. g. *De dynamidiis Galeni, Secreta Hippocratis* and many astrological tracts.

[139] Dissection of animals was practised at Salerno as early as the eleventh century.

[140] The sources of the anatomical knowledge of the Middle Ages are discussed in detail in the following works: R. R. von Töply, *Studien zur Geschichte der Anatomie im Mittelalter*, Vienna, 1898; K. Sudhoff, *Tradition und Naturbeobachtung*, Leipzig, 1907; and also numerous articles in the *Archiv für Geschichte der Medizin und Naturwissenschaften*; Charles Singer, 'A Study in Early Renaissance Anatomy', in *Studies in the History and Method of Science*, vol. i, Oxford, 1917.

[141] Benivieni's notes were published posthumously. Some of the spurious Greek works of the Hippocratic collection have also case notes.